CITY FARMING

A How-to Guide to
Growing Crops
and Raising Livestock
in Urban Spaces

CITY FARMING

A How-to Guide to Growing Crops and Raising Livestock in Urban Spaces

KARI SPENCER

First published 2017

Published by
5M Publishing Ltd,
Benchmark House,
8 Smithy Wood Drive,
Sheffield, S35 1QN, UK
Tel: +44 (0) 1234 81 81 80
www.5mpublishing.com

A Catalogue record for this book is available from the British Library

ISBN 9781910455906

Book design and layout by Alex Lazarou

Printed by Replika Press Pvt. Ltd, India

Photo credits
All photos by Kari Spencer except for:
Figures 1.1, 3.2, 3.4, 7.1, 7.2 and 8.1 by Dreamstime
Figure 4.6 by Laura Denyes and Wish We Had Acres
Figure 6.1 by Canva
Figure 6.5 by Jill Green and Sweet Life Garden

Contents

3
SOAK IT UP!
Watering to Promote Vigorous Plants, Healthy Animals and a Flourishing Environment, 73

4
A TIME FOR EVERYTHING
Gardening Through the Changing Seasons, 95

5
MAKE ROOM FOR ABUNDANCE
Utilizing Urban Spaces Efficiently to Grow and Do More, 121

Acknowledgements

Most of the photos in this book were taken with my own camera. However, two farms loaned photos to me from their own portfolios. I would like to acknowledge Laura Denyes and Wish We Had Acres for Figure 4.6, depicting two adorable goat kids. I would also like to acknowledge Heather Rodman of The Real Leopardstripes for Figure 8.6, depicting a small greenhouse. These photos appear by their kind permission, as well as additional photos from both of their farms that are posted on the *City Farming Book* website.

I would also like to thank all of the farmers who allowed us to visit and take photos of their farms. The time that all of you spent giving tours and answering my questions is greatly appreciated. And a big thank you to Emily and Lewis for accompanying me on farm tours and for taking many of the photos that appear in the book and on the website.

I will forever be grateful to my extraordinary husband Lewis and to all of my exceptional daughters who have made our urban farming projects so much fun, and who were supportive in so many ways during the writing of this book. Our family is the heart and soul of The Micro Farm Project and the key ingredient that makes it so special.

I also owe a debt of gratitude to Greg, from whom I have learned so much and who graciously took up slack for me this year to keep Grow PHX on track. Thanks for all of your *what ifs*.

This project would not have taken place without Sarah, Senior Commissioning Editor at 5m Publishing, who offered me the opportunity and encouraged me along the way. I am so appreciative of your confidence and enthusiasm for *City Farming*. It has been an honor to work with you and the outstanding team of people who made this book happen.

Foreword

We are in the midst of a revolution. A peaceful one that I have dreamed of being part of since the mid-1970s when in the 8th grade I wrote a paper on how we were over-fishing the oceans. To this day, I am not really clear about HOW I knew there was a problem, just that there was something inherently wrong with how we were living on and eating from the earth.

Our revolution is a discovery of where our food comes from and what is in it. A return to the good ol' days when we knew what a peach really tasted like because at its peak ripeness it dropped into our hands and we took a bite, the flavor exploding in our mouths and sending tingles to our toes. This is because we harvested it from a tree that we nurtured, adding only organic mulch, compost and natural fertilizers, making sure no dangerous chemicals were used that would travel into our food. We have grown a clean, healthy peach that tastes great, is nutrient dense and picked at its peak ripeness. This is the healthiest food we can find.

The good news is people of all ages are jumping on board and growing food in their front and backyards, on their windowsills and in community and school gardens. They are joining the trend by the millions in a discovery of what is best for their health, for the health of their community and for the world.

What is exciting for me is that this process is being led by a new breed of farmers that I affectionately call *urban farmers* … with growing spaces ranging in size from a few pots to dozens of acres – growing food within or adjacent to our cities. Over the past decade I have purposefully changed the conversation from the notion of being a "gardener" to one of being a "farmer" because gardening is thought of as a hobby and farming as a vocation. This change in perspective plants the seeds for people to begin the process of creating a career in their yards and at the same time supports and nurtures their local food system.

Becoming an urban farmer is as simple as shifting the way you think about growing food:

- First, grow food.
- Second, share it with someone, as this is what farmers do. Even if you are just sharing it with your family.
- Third, name your farm!

The first two steps are pretty straightforward but why the third? Naming your farm builds and nurtures the local food movement in ways that you may not imagine by activating the conversation around food and farming. Imagine sharing the name of your farm with your friends – perhaps Jack's Bean Stalk, Laura's Roost or Gloria's Golden Guild – learning the name of your farm brings a smile to most people's face and inspires the question: "You are a farmer? Tell me about that."

A few years ago Kari Spencer walked into the Urban Farm Nursery and we spent an hour or so chatting. I saw the spark in her eyes as she told me about her urban farm called 'The Micro Farm Project'. I decided then that I had to figure out how to work with her.

Our job as we create our urban farms is to build resiliency into the spaces where we work and live. Kari has a special talent for magically being able to integrate all the parts of her urban farm into one cohesive model, one that works with the flow of nature rather than against it. Her systems are able to bounce back in the face of adversity, truly demonstrating what resiliency is all about.

As you step into your urban farming adventure, learning the basic tenets of how to grow your own food and what components come together to create an urban farm are essential. Kari is brilliant when it comes to being able to articulate this. She has taken a topic that may seem new or challenging and has composted it down to simple steps that will help you navigate "inventing" your perfect farm. Her step-by-step, easy-to-understand process makes the discovery exciting and fun.

Because Kari has spent most of the last decade exploring and experimenting with these concepts, she has gained invaluable insights and lessons that she shares in *City Farming*. This book will give you the tools to create your urban farm and join the food growing revolution. I often say that things get done in the world because people take them on, committed to change their world. I challenge you to dream big, plan your farm and raise your first meal entirely from your yard.

GREG PETERSON
The Urban Farm
Phoenix Arizona
December 2016

How To Use This Book

I'm a city girl. I love the hustle-bustle, the entertainment choices and social activity, and having friends and neighbors within close proximity. But, I'm a little bit country, too. My inner farmer longs to get my hands in the dirt, to hear the early morning rooster crow, and to gather around the dinner table to enjoy the fruits of our labor.

I am not alone. A movement is growing, comprised of urbanites who would rather pick tomatoes than mow grass, and for whom chickens are more appealing pets than hamsters. Perhaps you are part of the movement. If you count yourself amongst those who live in the city but long for the fresh smell and feel of the soil; if you desire a creative connection to the earth and to the source of your food; if eating locally grown foods means stepping out of your back door to pick produce from your garden, you're in!

This book was written with you in mind, whether you have been growing or raising food for a while, or are just starting out. A *how-to* manual for gardening and raising livestock, the pages are packed with practical information. Each section spotlights examples that show how backyard farmers creatively apply agricultural *know how* to the real world of urban farming.

And although urban farmers employ a variety of conventional growing methods, there is a new movement amongst urban food producers that is far from conventional! It is revolutionizing what it means to be a *farmer*. Enter the exciting world of Permaculture, the art and science of working with nature to create landscapes that are increasingly productive over time with minimal labor and expense.

Learn how to garden, explore what it takes to keep chickens, goats and other animals, and discover specific urban farming techniques that range from simple and inexpensive to more complex for those who like a challenge!

City Farming will help you to understand the basic principles of Permaculture and successful urban farming. It will provide you with an arsenal of practical tools and techniques that you can apply to produce bountiful food on your own property. Best of all, you will be equipped to come up with your own innovative solutions to create the urban farm that you desire.

CHAPTERS

Chapters are divided according to specific urban farming challenges related to growing food in the city. Chapters are organized in such a way that after your initial reading, you can easily flip through to find the solutions that apply to your individual needs. Whether you are looking for ways to grow a garden in your specific climate, want to explore space-saving techniques, or need answers to questions about livestock, soil, pests and other topics, chapter titles and headings will help you to find the relevant information quickly. Feel free to highlight text, dog-ear pages and use bookmarks to reference key passages.

EXERCISES

Chapters include helpful exercises entitled *Inventing Your Farm*. Whether you are already an urban farmer, or are exploring the idea, the exercises will assist you in creating, expanding or redesigning your homestead.

OPTIONAL SUPPLIES

To accompany the book and the website, you may wish to have the following:

- a notebook,
- a simple, hand drawn aerial-view map of your property,
- several sheets of tracing paper.

These will be useful tools, should you wish to perform the *Inventing Your Farm* exercises suggested in the chapters.

WEBSITE

You will often see links to the CityFarmingBook.com website that you can visit for more information, virtual farm tours, interviews, photos, tutorials and tips. There are also links to the best online information and helpful resources for urban farmers. These links will help you to discover local supports and information relevant to your location and climate.

To access the information on the website, type the link address into your browser. With every link, you'll also see a bar code. It's called a QR code, and it's used as a short-cut to link to content online using the camera on your phone or tablet, saving you from typing lengthy addresses into your mobile browser.

You'll need an app that can read QR codes, which you can find in the app store on your device. When you see a QR code on a page, scan it with the barcode reader and it will automatically take you to the online content. Scanning Figure QR 0.1 will take you to CityFarmingBook.com. It's that simple. Go ahead and try it. Farm out!

Preface

"Mom!" came the distant but urgent call in the early morning darkness. "Hurry!"

My heart skipped a beat, adrenaline clearing my sleepy head as I raced to the door that lead out to our backyard farm. Early morning chores are typically peaceful, but today the tone of my daughter Emily's voice was uncharacteristically urgent. As a fledgling farmer, I wondered what could have gone wrong, and if I would be able to fix it.

Running to the sound, my daughter met me halfway, breathlessly stammering the words *Jane* and *baby goat*. With trepidation, I approached the kidding pen where my favorite doe, Calamity Jane, was secluded as she approached her due date. By the red light of a heat lamp, I could see a tiny, wet newborn lying still on the ground. My heart sank as my eyes searched the pen with alarm. We have had lots of ups and downs on the farm, and at that moment, I wasn't sure which this was going to be. Figure QR 0.2 will take you to the story of one of our most difficult days.

A flash of movement drew my gaze, and I breathed a sigh of relief as I spotted two shivering kids huddled together under the warmth of the lamp. Jane turned and began to nudge and lick the newborn on the ground. Resisting the urge to rush into the pen to assist – and not knowing exactly what to do – I held on to Emily's arm as together in the pre-dawn darkness we watched and waited. Moments passed, and finally a tiny squeak met our ears as the baby struggled to its feet. Putting my arm around Emily in a rush of relief and elation, I whispered "triplets" with a smile.

This was the yin and the yang of our first adventures into farming: bright promise balanced with constant vulnerability created by our own lack of preparation, education and experience. Perhaps you can relate. Have you ever jumped into a project with exuberance, excited about the possibilities but knowing in the back of your mind that you were a bit unprepared? That's exactly how we started out as urban farmers.

Trial and error are wonderful teachers. I am grateful for this truth because our first flailing forays into urban farming had no guidebook that we could follow on our mission to grow and raise as much food as we could on a city lot. Soaring to the heights when new babies were born or hatched, and when bumper harvests emerged, we were truly having the times of our lives on a blind adventure. Even so, our operation's vulnerabilities frequently surfaced, usually related to climate, faulty management and unexpected expenses. Our family quickly learned to take it in stride when crops failed or livestock faltered, chalking it up to the price of our farm education. Over time, things began to come together.

Word of our little farm began to spread, and it needed a name that people could recognize. We had considered titles that included the words *oasis* and *utopia,* but felt that these romantic terms would obscure the sweat and tears on which our desert farm was built. Right from the start, we knew in our hearts that we were not creating a place of retreat, but an ongoing project that would shape our future, a living laboratory that we named *The Micro Farm Project.*

Though we started off alone, today we are in good company. A swell of interest in urban farming has induced the establishment of many new city farms. And greater acceptance of growing and raising food in populated areas has brought undercover backyard farmers out of the shadows. Together, the seasoned veterans and the freshman farmers are rapidly changing how food is produced and accessed in our cities. Their influence is not limited to food production; urban farmers are creating solutions that meet some of the most pressing challenges of our day, from providing nutritious food for our families to affecting climate change.

When I consider how far we have come from our early days and the small part that we play in a much bigger urban farming revolution, it strikes me that there is no better lifestyle. It is my heart's desire to share it with others. That's why I wrote *City Farming*, to spread the word about the benefits of urban farming and to help people join in.

I am so glad that you opened this book. *City Farming* was written with specific folks in mind, and it is for you if …

- You are curious about urban farming and want to find out more about it.
- You are ready to join in as an urban farmer, and need information to get started.
- You have tried to grow food, with limited or dismal results, and want to improve.

- You need to figure out how to get around a hindrance to producing your own food.
- You are a seasoned urban farmer, and want to discover new methods to make it even better!

City Farming is both the personal story of our farm and a compilation of anecdotes from a variety of urban farms around the country. It contains some of the best urban farming solutions in practice today and is the guidebook that we wish we had in our hands when we began our farming adventure. Its pages illustrate the basics of how city farms produce food in the city. It also dives deeper, exploring ways that urban farmers are overcoming challenges, both big and small, to feed themselves and their communities, to nurture their neighborhoods and enhance the environment. Here's what it's all about:

- I'll show you the best gardening principles to grow abundant food on your property, whether your space is big or small.
- You'll discover the miracle of Permaculture, which is all about working with, instead of against, nature. It's a fascinating and proven way to get the absolute most out of your garden and your property – at the smallest cost and for the least amount of work.
- You'll learn how to raise the happiest, healthiest livestock that produces the highest quality food – and how animals can do a lot of your farm work for you, too!
- I'll share design tips and action steps to take your urban farm from dream to reality. By following the action steps, you will create a farm that maximizes all of the plusses of homesteading – abundance, enjoyment and health – while minimizing or eliminating the minuses.
- Each chapter contains specific, practical examples and methods employed by urban farmers, as well as *how-to* information that you can apply to optimize your own urban farm.

Urban farms have challenges, from small spaces and poor soil to city ordinances and climate extremes. Within every roadblock to growing food, there lies a solution. Often the solution is simple and easy to achieve. But the truly exciting possibilities emerge when we combine knowledge with creativity to turn existing problems into *valuable assets* for ourselves and our communities.

Are you ready to have an urban farming adventure and to change your world?

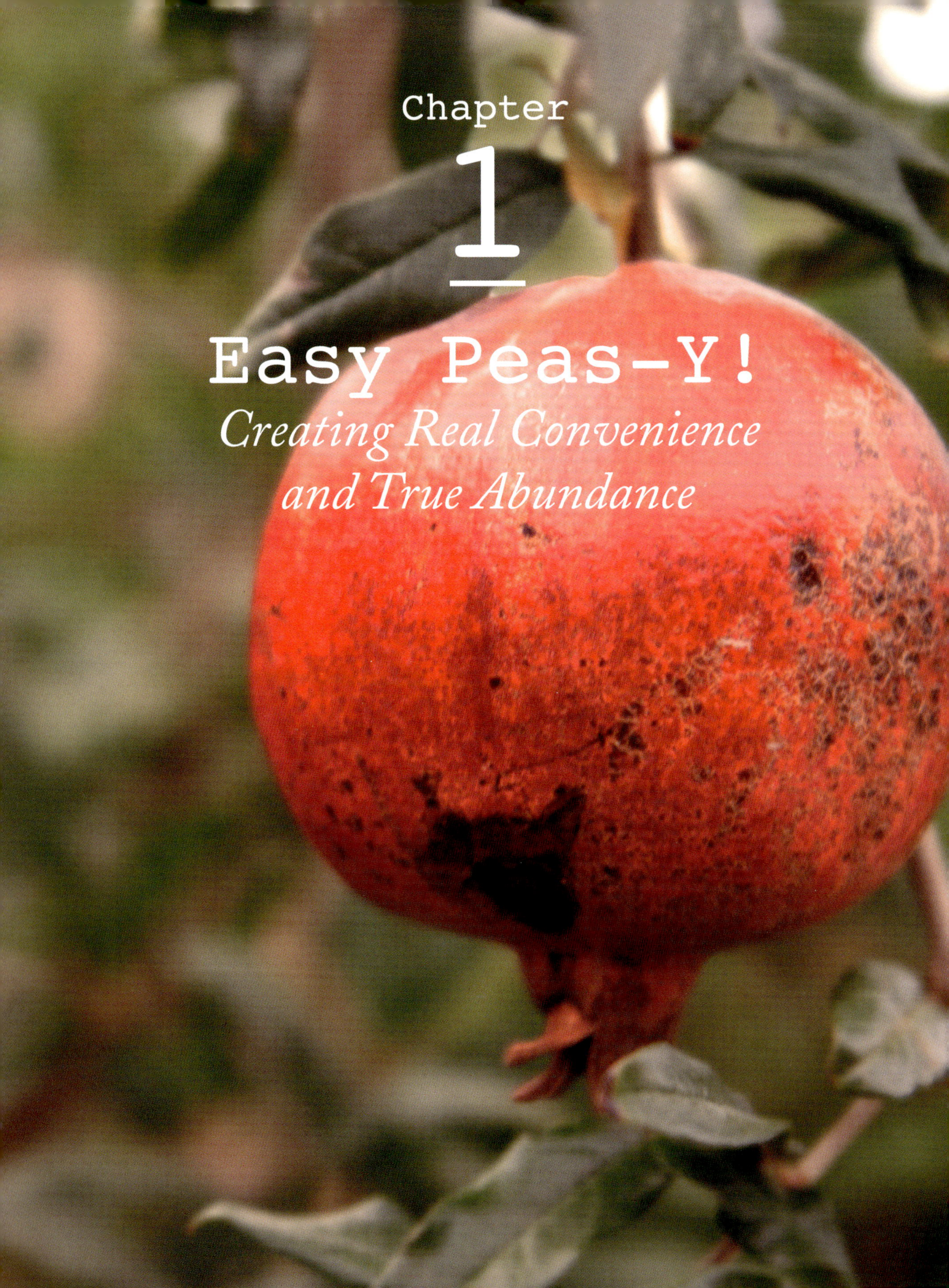

Chapter

1

Easy Peas-Y!

Creating Real Convenience and True Abundance

Making a getaway in the minivan, I nibbled uneasily at my hamburger and considered that things had not gone exactly as planned at the convenience store. The day had started as usual, haphazardly, three small nightgowns strewn on the floor in exchange for jeans and boots, quick gathering of milking supplies and out the door to the farm. We had slept late, relishing the last few days of winter break, and the animals were annoyed, impatiently awaiting their breakfast. I knew the neighbors would be annoyed, too. I prodded the girls to hurry up with the feed to halt the hungry complaints, the cacophony of clucking, bleating, gobbling and periodic clanging as Oliver the ram expressed his displeasure at our tardiness by colliding with the metal fence head-on.

The girls scattered to their tasks as I settled into ritualized goat milking, my mind free to wander as the creamy liquid whooshed rhythmically into a pail. Gazing on the makeshift pens scattered slapdash across our small property, I recognized that we had all of the symptoms of *urban farming fever.* We were collecting animals with something akin to the fervor of Noah, not to save ourselves from impending doom, but from a slow death at the hands of the conventional food system.

There had been simpler times when the yard had just been a patch of dirt, back when I was blissfully unaware, happily living on convenience foods and congratulating myself on how cheaply I was able to feed our family of six on a narrow budget. Life was comfortable, but change was brewing. Between 2006 and 2010, our oldest daughter went away to attend university and my husband was deployed twice by the United States Navy Reserve. I was left alone with three young children and a serious need for diversion from the lonely stretches of time that stagnated like mud in a dry river bed.

The idea began to grow in my mind to transform our dirt backyard into a garden. This budding desire was based on nostalgic remembrances of my grandfather's garden and how much I had loved searching out ripened peas, pulling carrot tops to reveal a bouquet of sweetly crisp roots, or slyly plucking raspberries to savor in solitude. I broke ground with the hope of ushering the simple joy of my childhood memories into our present reality.

And so, what would eventually become our farm began with a spontaneous hobby garden, intended purely for pleasure. But as I rubbed shoulders with other growers and farmers, my eyes were opened to a whole world of troubling facts that I couldn't ignore.

I began to recognize that the conventional food system was not focused, as I was, on the joy of growing superior, delicious food, but almost exclusively on increasing efficiency. The concept of *efficiency* had never held for me a negative connotation; that is, until I connected it with food production and discovered that the goal of this efficiency was not to create tastier, healthier foods, but with a dual aim to lower consumer costs and maximize overall production. In other words, cheap, abundant food was paramount, generally at the expense of flavor, nutrition and any hint of artfulness.

I remember the day when I gazed at a stack of tomatoes on the store shelf, every one of them perfectly plump and uniformly red. It was suddenly obvious to me that they had been selected for shelf-life and appearance rather than flavor. Estimating that the tomatoes had been picked weeks ago, I wondered how much nutrition had been lost in transport. It dawned on me that the reason the tomatoes at the grocery store did not taste as good as those that were home-grown in my garden was not a fluke or a product of any superior tomato growing skills that I might possess, but the reality that the tomatoes in the store had been cultivated for durability, picked early and ripened in transport. This was a far cry from the tomatoes ripened to maturity on the vine in my garden, selected for flavor and unique coloration, an array of yellows, oranges, reds and purples.

Now that my eyes had been opened, I saw the consequences of society's collective drive for efficiency everywhere in the market, from vegetables to meats and prepackaged foods. My intentions for our farm now became twofold: to escape the trappings of the industrialized food system and to embrace the ideal of a natural, artful and simple farm life. Simplicity, however, does not always equate with ease. The rewards of farm life are reaped with the hard work of consistency, planning and the ability to stick to a course of action.

I jumped into urban farming full throttle, as yet unaware that I was overestimating my ability to focus on a course of action, and that I was ignoring basic facts about my personality. Looking back, I am sure that I had romanticized the farm life in my imagination, deluding myself into believing that I could settle easily into the routines, contentedly passing the days in overalls, without interruption, fully focused, living the life pastoral.

Having studied myself over a number of years, I have discovered that consistency is not my strong suit. I think that if my biography was written as a poem, it would be a collection of haikus – each precise, poignant and *brief*. I can do anything excellently

and with great motivation … once. Sometimes my enthusiasm lingers for weeks, even months. But the prospect of doing much of anything over and over again in perpetuity is unfathomable to me. That is, of course, unless I see a clear and powerful reason for it, as in the case of brushing my teeth or going to work, which are rarely ever compromised, even on the most distracting or stressful of days.

And so, I had a sense that I *should* grow as much food as possible, and what I couldn't grow myself, I *should* purchase from organic sources. But these ambitious *should(s)* did not yet rise to the level of *no compromise*. I would describe my behavior at the time as desiring to grow and eat healthy foods. Yet, if the day was stressful or other things came up, I was quick to order a pizza. In hindsight, my intentions had been based on academic knowledge of the shortcomings in our food system, and a rueful sense that I *should* do something to change it. But lacking any urgency or emotional punch, neither of these motivations was compelling enough to overcome my distracted nature, nor to forestall the events that were about to unfold before me that day at the convenience store.

My mind returned to the tasks at hand. Hurriedly washing the milk bucket at the kitchen sink, I checked the time on the oven clock and called for the girls to finish up and pile into the van. Sleeping in had given us a late start, and we were short on time to make it to the junior sheep judging contest at the National Livestock Show. I wanted to check out the sheep breeds and hoped to spark a desire in my kids to join 4-H. We needed gasoline and a quick bite to eat on the way, and I hadn't budgeted any time for either.

Against my better judgement, I pulled into the convenience store and parked at a pump near the door that gave me a bird's eye view of the shop's interior. Handing my twelve-year-old a twenty-dollar bill, I sent the girls inside with instructions for each to get a hot dog or a burger, a bag of chips and a drink. I pumped the gas, watching them as they milled around the store, looking at all of the choices on the shelves. I had hoped that they would have completed their purchases before I finished with the gas, but had no such luck.

I pulled the van into a parking spot and entered the store to round up the kids. In the face of so many choices and limited funds, they were having trouble deciding who would share with whom so that they could each get a full meal and have money left over for sweets. The negotiations started with two requesting to share a hotdog and chips in hopes of also getting ice cream and another pleading to upgrade to a shaved ice instead of a soda. I foolishly granted permission, at which time the debate turned to which brand of chips the two would share, and how it wasn't fair that the other got a shaved ice when they did not. Walking away from the squabble, I perused the aisles in search of something for myself. Grabbing a foil-wrapped burger from under the heat lamp, I called for the girls and shuttled them to the counter. Resignedly adding another five-dollar bill to the twenty that I had given the kids, we quickly gathered our purchases and jumped into the van.

Unwrapping my burger with one hand and steering into traffic with the other, I left the unfortunate scene behind, wondering how much sticky, melted ice cream, grease and brightly pigmented liquids were being smeared around the back of my van. *Good grief,* how would the dirtied upholstery smell in a couple of days? The thought grew even more disturbing as I considered that these disgusting items were going into my children's bodies; that I was feeding them something that was not quite food and likely somewhat toxic. Furthermore, I had wasted twenty precious minutes and paid $25 for the privilege.

Righteous anger rose up inside of me, and regret. My stomach twisted as I thought of the purple lettuce waiting in the garden to be picked, but ignored. How easy it would have been to pack salads for the road. Had we made sandwiches, even with pricey organic bread and cheese, it would have cost less, tasted better and certainly been more nutritious than what we just ingested. Instead, those healthy items languished at home while we squandered our time, money and health on convenience foods.

What on earth was convenient about that? My hands gripped the wheel as tightly as my mind gripped on to resolve. I was angry. I had been tricked, not just today at the convenience store, but over and over again. Somewhere along the way, I had allowed so-called convenience foods to be elevated from occasional options to be used in a pinch to the predominant food source for our family. From restaurant meals to frozen entrees and prepackaged goods, images of my culinary habits rose up and swirled like a cyclone in my mind. How many minutes had I spent waiting in drive-through lines longer than it would have taken to put a fresh meal together at home? How many times had I run to the store for an onion when I could have planted hundreds of them in just a few hours? How often had I paid four or five dollars for an ounce of fresh herbs when I could have grown rows and rows of them in my yard for less money? How many hours had I spent wheeling up and down the aisles for canned vegetables and frozen fruits when I could have preserved home-grown produce in my own kitchen in a day and have it at my disposal all year long?

I suppose that the combination of marketing and my own fly-by-night lifestyle had made me a willing dupe for the industrial food system in spite of the superior food that was being produced in my own backyard. I couldn't change the system, but I could change my response to it. No longer viewing it as a friend, I had an urgency to not only grow food, but to grow the most food and the best food possible.

And then, as obvious as it might sound, to actually *eat* that food and not let any of it go to waste.

Knowing my own shortcomings and the potential for me to lose focus in my gardening efforts, I had to figure out how to make home-grown foods as convenient as possible – more convenient than convenience foods – so convenient that it would be a hassle *not* to eat from our own food stores.

Over the years, I have grown more proficient at using the harvest to its full potential. And I am moving closer to becoming a *Growitarian*. Figure 1.1 describes the *Growitarian* style of eating that I am striving to adopt. The process starts before the

1.1

Growitarianism

Broadly defined, a growitarian is a person who eats mainly personally grown fruits, vegetables, nuts, legumes, grains, seeds, eggs, dairy, meat. Many growitarians eat foods grown by local producers, but avoid food produced by the industrial food system.

first seed is sown, with planning and a good bit of dreaming about what will soon become a reality in my garden. During the planning stage, I decide what to grow and how much. Then I calculate when those items should be coming into their peak of ripeness and formulate a plan of attack to use every bit of the produce.

This chapter will focus on the planning, dreaming and preparation that will create extreme abundance and simple convenience on our pantry shelves. It will help us to develop a clear plan for how we will use the harvest when it arrives, so that none of it may go to waste. Though the rest of this book will highlight techniques to make *growing* food as easy and successful as possible, this chapter skips ahead to ensure that we will make use of the incredible harvest as it ensues.

WHY GROW AND RAISE FOOD?

When I have taught this topic in live classes, I have always saved it for the last week of instruction. But as I thought about the logical layout of this book, it occurred to me that this information ought to come first. One of the most discouraging things that I occasionally hear from former students is that they have stopped gardening because they just weren't using what they grew. To prevent such loss, it is vitally important to know *why* we are producing home-grown food and specifically *how* we will incorporate it into our culinary lifestyle before we jump into technical aspects of growing and raising food. If we do not first understand why we are doing the work to produce our own food, when the going gets tough, it will be easier to give up than it will be if we have a clear motivation to press on.

The question is, *why do you want to produce your own food*? Perhaps for one of the following reasons:

- Self-sufficiency in a dependent world
- Solitude in an overly-connected world
- Security in a society that promotes the illusion of security
- Exercise and doing something worthwhile in a frivolous and sedentary culture
- Understanding the effort required to produce sustenance in a system where food is fast and cheap
- Connecting with nature on a regular basis, not just on vacation
- Embracing simplicity and a connection with the past

- Doing something meaningful as a family or with others in your community
- Eating better tasting, more nutritious foods
- Eliminating toxins from your family's diet
- Having home-grown and home-produced foods available, right at your fingertips
- Eating locally produced foods (what's more *local* than your own backyard?).

These are a few of the reasons that I grow food. Yours may differ from mine. Whatever your motivations, take a moment to write them down. This will make them more concrete and harder to discard when you are going through the growing pains of starting your farm. If needed, tape the list to the refrigerator door or bathroom mirror to remind yourself *why* you are going to all of the effort to produce your own food. And if you have a partner or children who will be part of the effort, prepare your list of motivations together.

INVENTING YOUR FARM

What are your motivations for starting an urban farm? Write them out on a sheet of paper or in your garden notebook. First, brainstorm all the reasons that you want to be an urban farmer and jot them down. Then, number them in order of importance. Talk it through with your family or farm co-creators.

I highly recommend that you name your farm and use your list of motivations to create a mission statement. A name and a mission statement will give your urban farm credibility and legitimacy, a sense that it is a bona fide farm and not a fly-by-night, makeshift operation. Even if you do this only for yourself and not for public consumption, having a name and a mission statement will empower you to clarify and fulfill your urban farming dreams.

I have given a lot of thought to why our farm exists and why we want to grow our own food. Logically, it just makes sense to grow food for the health and economic benefits. Beyond the facts, growing food touches me on a more spiritual level, satisfying parts of my soul like nothing else. Though I am not a poet, Figure 1.2 expresses the emotional reasons that motivate me.

1.2

HARVEST PLANNING

DECIDING WHAT TO GROW

Now that you have clarified *why* you want to grow, you can determine *what* to grow. Start very simply with a few things that you like to eat, especially those items for which you often make emergency runs to the supermarket. At The Micro Farm Project, I started with a few herbs, garlic and onions. Fresh herbs are expensive to buy and have a very short shelf-life, so having them on hand immediately made cooking with them more convenient and less expensive. Eggs were another item that I used daily, so we started our livestock production with chickens.

When making your list, consider how much space you have in your garden. Mature vegetables come in all sizes, from Globe basil that grows in a compact mound to sprawling watermelon vines and parasol-sized zucchini leaves. Livestock can take up an enormous amount of space, creating the situation in which a backyard farmer may have to choose between having a large garden, or a smaller garden to make room for animals. Allocate the space available to the foods that you desire the most.

Finally, show your list to experienced gardeners in your area to determine whether they are in season and how challenging they are to grow in your area. Start with the easiest varieties to grow, and save the more difficult varieties to try as you gain confidence in your skills.

DECIDING HOW MUCH TO GROW

I am often asked how much of any given vegetable a person needs to grow or how many animals they will need to raise in order to feed a family. The query is often framed in terms of raising enough food for an entire year. Although the goal is admirable, a year's worth of food is a tremendous amount to grow. And it can be daunting to figure out just how much you should plant. For this reason, in the beginning stages of your urban farm, I recommend cultivating a small practice garden, and worrying about calculating yields later.

There is no need nor expectation to become self-sufficient overnight. Your first growing season is for practice and observation. Keep track of how much you planted, how well the items grew, and how many pounds of food the plants produced. My first season growing green beans, I had no earthly idea what my yields would look like. I planted a small three-foot patch. The beans grew beautifully, but it would take several days for me to harvest enough beans for a meal. That season, I did not calculate the bean harvest in pounds, but in the small number of meals it provided. The experiment, while not high yielding, was by no means a failure. I learned that beans are easy to grow in my climate and that I needed to increase the size of my patch. With experience under my belt, I was better equipped to estimate how many seeds I would need to plant in order to have enough beans for fresh meals and some extra for canning the next season.

For those who prefer a more exact answer to the question of how much to plant, the answer is that *it depends*. Family size, preferences and eating habits differ. And estimated plant yields can vary by region. Even so, online crop calculators can be helpful to provide a basic idea of how many seeds to plant, how much you will need to grow for your family size, and yield estimates. Figure QR 1.1 will take you to more information about calculating crops.

As you cultivate your practice garden and become a more proficient urban farmer, you can make adjustments. Keep records of how much you plant and how much your garden yields in terms of pounds, quarts or the number of meals the harvest provides. Do the same for your animals, journaling the number of eggs, pounds of meat, or quarts of milk that they provide. This information will help you formulate a better understanding of your family's needs and to hone your operation towards greater self-sufficiency.

KEEPING TRACK

Keeping records of farm activities may not seem very exciting, and it is not my strong suit. But I have found that when I keep good records, I become a better urban farmer. Here are a few record-keeping tools and how I use them:

Journals

I journal about all kinds of things that occur in the garden. Although journaling may be inconvenient in the short term, having a garden journal definitely makes urban farming easier down the road. I was terrible at keeping a farm diary at first. But several instances occurred in which a recurring problem arose, and I could not remember when it had appeared the previous year or exactly what we had done about it. Instead

of continually trying to rack my brain for information, I finally decided to be diligent at writing things down.

In the interest of time, I keep a simple pen-and-pencil diary consisting of brief notes. If a pest or disease issue arises, I document it, logging when it occurred, what I did about it, and the result. When I try a new farming technique or make a change of any kind, I make notes about how well or how poorly it went. I also jot down interesting occurrences or things that I want to research. Looking over my journal through the years has been very helpful for me to spot patterns and to learn from both my successes and mistakes.

I also keep a file on my computer to store information about our animals. One file keeps track of baby goats and sheep that are born on the property, noting dates of birth, litter size, the sire and dam, health and vaccinations, and to whom they were sold.

There are numerous ways to journal. I recently started using a voice recorder to quickly note items that I want to add to my garden diary. And I know several gardeners who make entries using their smart phones or who carry an iPad with them to the garden. One of the benefits of electronic journaling is search-ability. As an example, if you are having a recurring pest problem, you can search the journal document to find all of your previous references to the problem. This can be extremely helpful to spot recurring cycles and to understand what works and what doesn't on your particular farm.

Calendars

In conjunction with my journal, I also keep a garden calendar. On the calendar, I note the dates on which I plant seeds and also when I should start looking for a harvest. For example, when I set a tomato transplant that takes ninety days to mature, I mark both the planting date and the expected harvest date ninety days later. I also keep records of exact harvest dates, noting when I harvested and how much.

A second way that I use my garden calendar is for farm planning. I can put a reminder on my calendar to plant items at a particular time, and to plan ahead so that I am not away on business or vacation when the produce is at its peak of ripeness. I also use the calendar to time animal husbandry so that babies are not expected when the weather is too hot, too cold or while we are planning to be away from the farm.

Finally, I use the calendar to track rainfall and weather patterns. This has been extremely useful over the years in helping to anticipate weather events and to better understand the seasonal changes that occur on our farm.

Plant Labels

When planting, identify your crops with plant labels. This may sound like a *no-brainer,* but I have very often been absolutely certain that I would remember what I planted, only to scratch my head later and wonder what it was. I once received an heirloom tomato start from my friend Suzanne Vilardi, the proprietor of Vilardi Gardens. For one reason or another, I did not label it. The plant turned out to be vigorous, producing delicious tomatoes that were perfect for sauce. The next season, having forgotten the name of the variety and wanting to replace it, I attempted to describe it to Suzanne. She laughed and told me that she grew several tomatoes that could fit that description. Fortunately, as I perused her inventory, I happened upon the Punta Banda tomato, the very one that I had grown the year prior. Happily, I took the Punta Banda start home, making certain to label it and jot down a note on my calendar, too.

Labels do not need to be sophisticated or artsy. They can be made from any number of materials, from tongue depressors or wooden stakes to commercial label sticks. If you prefer more imaginative labels, there are hundreds of creative and artistic plant label ideas on Pinterest and other websites. Use acrylic paint or a black permanent marker for writing the information on the label; anything else will fade in the sun and become illegible. Figure 1.3 shows how The Simple Farm in Scottsdale, Arizona, uses chalk paint and chalk markers to create lovely and long-lasting labels.

1.3

PLANT LABELS

PLANT NAME & PLANTING DATE

YOU MAY ALSO WANT TO POST THE DAYS TO GERMINATE, DAYS TO HARVEST & THE PLANT'S LATIN NAME

Mark the name of the crop, the planting date, and how many days to the harvest (which we will learn more about in Chapter 7, concerning propagation.) When planting seeds, you may also want to include how many days it should take for germination. This information should be available to you on the seed packet, and it is very useful for someone as forgetful as I am (as established by the Punta Banda tomato incident.) Figure QR 1.2 will take you to more information about plant labels. This year, I planted lemon balm and thyme. Both take roughly eleven to fourteen days to germinate. The seeds have not yet come up. I failed to jot the planting date on the labels, and, even in the short amount of time since planting, I do not remember on what day I planted these particular herbs. Therefore, I am unable to determine whether or not the expected germination period has passed, which would tell me whether or not I need to replant with fresh seed. And so, I am stuck waiting for a few more days to see what happens, just to be sure that I have given the seeds enough time to sprout. Next time, I will take my own advice and jot down the planting date.

Finally, I would encourage writing down the botanical name of the plant, either on your label or on your calendar, or both. Many plants that are completely different varieties share the same common name, as is the case with chamomile. Typically grown in home gardens are Matricaria chamomilla (aka German chamomile) and Chamaemelum nobile (aka English, Roman or Garden chamomile.) Though these plants have similar looking flowers, they are very different. German chamomile is a short-lived plant that grows to a height of 2ft. Roman chamomile is an evergreen that lives for many years, is low-growing and spreads like a ground cover. Knowing which specific variety you have will help you to either replace it with the exact same plant, should the need arise, or to avoid it altogether if you were not happy with it.

QR 1.2

Garden Maps

Our youngest daughter, Emily, is very good at making garden maps. When we plant, she will draw out the garden plot and record what is growing in each location. This is extremely helpful for two important reasons. First, if our plant labels fade or fail, we have a backup record. Secondly, it helps us to plan out our plantings before we sow any seed. Emily does the planning in pencil, in case things change as we are working, which is prone to happen. But even if the plan changes as we go along, it is helpful to

have an idea of where to place the seeds and transplants in advance to make the most efficient use of space.

These are the ways that records are kept at The Micro Farm Project, but it is certainly not an exhaustive list of methods. You many come up with a way that works better for you, or you may choose to discard one particular method in favor of another. Do that which seems most useful to you. What I can assure you is that, no matter what system you use, putting a little bit of effort into a record-keeping routine makes urban farming so much easier in the long run.

PREPARING FOR THE HARVEST

When crops mature, it is important to harvest and use the produce quickly. Vegetables not only taste best when they are at their peak of ripeness, but they are also at the pinnacle of their nutritional density. Once picked, fruits and vegetables immediately begin to lose flavor and nutritional value. Vitamin loss can occur rapidly, as in the case of spinach, which has been shown to lose up to 90% of its vitamin C within three days. [Favell, D. J. (1998) A Comparison of the Vitamin C Content of Fresh and Frozen Vegetables. *Food Chemistry* 62.1, 59–64.]

Vitamin and flavour loss can be slowed from a sprint to a crawl by swiftly eating or preserving harvested vegetables, fruits and animal products. In general, food items gathered from your farm should be quickly chilled and prepared. Valuable time and nutrition can be lost in searching for recipes and gathering supplies after the fact, so here is what I recommend to maximize the harvest:

- Mark on your garden calendar the timeframe in which you expect the harvest to come in. If you are growing enough to preserve, try not to overbook your schedule during this period as preserving the harvest in a timely manner will require your time and attention. Because I really love to preserve tomatoes, I try not to schedule vacations, business trips, or big events when they are in season.
- Collect recipes in advance. As the harvest approaches, gather a handful of fresh recipes for quick, healthy meals. Also, decide how you will preserve the excess. If you will be making pickles, relishes, jams or other preserves, have those recipes at your fingertips, as well.

- Discover the many ways to preserve foods. If you are unfamiliar with food preservation techniques, take the time between planting and harvest to discover your options and decide which methods most interest you. Later in this chapter, we will cover a few of the techniques that I have used at The Micro Farm Project.
- Collect equipment and supplies as the harvest approaches. Preservation methods often require special tools and materials. Don't wait until the last minute to acquire them. By purchasing supplies in advance, you can search around for the best deals, or even find a friend who will loan them to you. When I first attempted dehydrating foods, I borrowed a dehydrator to test it out before I spent the money on my own, and then I perused kitchen resale stores until I found the style I wanted at a discounted price. Likewise, shopping for small materials, such as canning jars, freezer bags or parchment paper in advance saves you an emergency trip to the store on harvest day, and allows you to buy them as economically as possible.

  The same is true for animal products. When a goat is about to kid and come into milk, I gather milking supplies and canning jars. And during the winter when my chickens halt laying, I collect used egg cartons from friends so that I will have enough to get us through the summer when we will have a dozen or more fresh eggs each day.
- Place equipment within eyesight. Over the years, I have collected a lot of kitchen equipment, from canners to strainers and food processors. Much of it is bulky, and used infrequently, so I store it away in the garage or up on the highest shelves of my cabinets. When the harvest approaches, I bring it into the kitchen, dust it off and keep it handy so that it is readily available when I need it. This simple step saves me a lot of stress on harvest days when there are so many other things to do. And having the equipment within eyesight gets me excited about the harvest and reminds me to check my garden for signs of ripeness.
- Collect shelf-stable ingredients. Once I have my recipes in hand, I can begin to collect the spices, pectin, herbs, and other dry ingredients that I will need to prepare them. I also stock up on side items, such as pasta, quinoa, rice and beans that will complement the recipes. This makes it very easy to prepare fresh meals or preserve produce quickly without having to run to the store at the last minute.

Having recipes and supplies handy takes the stress out of harvesting, and always gets me very excited about the potential for fresh meals on our table and delicious preserves on my pantry shelves. It prevents the inertia that I often feel when I am unprepared for the harvest. And it arrests the danger of wasted produce that occurs when I allow myself to become overwhelmed by the process. Just a little bit of advance planning and preparation ensures that every bit of our produce gets used and that we have fresh foods on the shelves and in the freezer every day of the year.

FOOD PRESERVATION METHODS

Preserving food is easy to do and the benefits of having home-produced foods on the shelves far outweighs the effort that it requires. Preserving can also be an enjoyable and creative process, giving you the satisfaction of creating healthy and unique food items that money just cannot buy.

Nothing is better than needing a jar of tomato sauce and getting it from my own pantry. Or sending my kids to school with homemade applesauce, fruit leather or candied pineapple for their snacks. Or serving goat cheese made with raw milk and home-grown herbs at a party. Or giving homemade preserves to the kids' teachers at the holidays. The list goes on and on, and my goal is to minimize trips to the store in favor of having abundance at home. Here are some of my favorite food preservation methods:

WATER BATH CANNING

I love a well-stocked pantry filled with home-canned goods, and water bath canning is one of the easiest ways to achieve it. At the end of a day of canning, I am tired, but triumphant, savoring the work of my hands and the knowledge that my family will enjoy the flavors of summer all winter long.

In addition to preserving excess harvest for the leaner gardening months, canning has a host of benefits. I enjoy the convenience of having ingredients on hand so that I don't have to run to the store for staple items, such as tomato paste and diced tomatoes. By canning my own goods, I control the ingredients and know exactly what is in the foods that we are eating. Canning also stretches our budget, preserving items when they are in season to use when the price at the market goes up in the winter. Figure QR 1.3 will take you to a detailed article that explains the process of canning tomatoes, using a water bath canner.

Water bath canning is a method of preserving foods in glass jars using boiling water. It is ideal for high acid foods, such as fruits, jams, salsas, condiments and pickles, as the acidity prevents harmful bacterial growth. Properly sealed food jars will last on the pantry shelves without refrigeration for months, or even years, without spoiling.

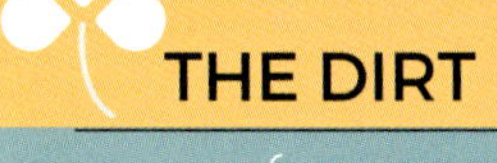

One of the things that I love about canning is that supplies are inexpensive and reusable. To get started, you will need to acquire a canner, which generally consists of a large enameled pot with a removable jar rack. Other equipment includes a jar lifter for removing hot jars from the boiling water, a funnel for filling the jars, and the jars themselves. Everything but the self-sealing lids can be used over and over again.

THE DIRT

on farming

Home Food Preservation

CITYFARMINGBOOK.COM/HOME-FOOD-PRESERVATION/

QR 1.4

Recipes abound online, in current canning books and magazines, as well as in vintage cookbooks. When using vintage recipes or those created by online bloggers, make sure that the procedures comply with current USDA guidelines for safe canning published in *The Complete Guide to Home Canning*, which can be downloaded at ucfoodsafety.ucdavis.edu/files/26457.pdf. Figure QR 1.4 takes you to more information and links about canning.

PICKLES AND RELISHES

Pickling is the process of preserving or expanding the lifespan of food by either anaerobic fermentation in brine or immersion in vinegar. You can pickle most types of vegetables and many fruits, too. Just cut them into small enough pieces to fit into a jar, and then add a few seasonings and spices. Figure 1.4 shows how we pack pickle jars at the farm. Beyond packing the jars, there are three ways to preserve your pickles or relishes. The easiest method is to make quick pickles that are stored in a vinegar brine in the fridge. No further processing is necessary, and the pickles will last for several weeks.

1.4

PICKLES

THE MOST DELICIOUS PROBIOTICS

PACK CUCUMBERS TIGHTLY IN THE JAR, 3/4 FULL, OR THEY WILL SHRINK DURING PROCESSING AND FLOAT TO THE TOP.

If you prefer to have shelf-stable pickles, you can hot pack them and process them in a water bath canner. Hot pack pickling takes very little time and equipment. For the crispest pickles, soak the cucumbers whole in cold water and pickling lime overnight. Rinse well to remove excess lime. The following are the basic hot-pack pickling steps. Exact recipes can be found online or in any number of canning cookbooks.

Step One: Pack your Jars

Fill your jars with your choice of vegetables and seasonings. The jar should be about three-quarters full of vegetables, with just a few of your spices and seasonings mixed in or placed around the vegetables against the inside walls of the jar.

Step Two: Prepare your Brine

In a large pot over medium heat, combine water, vinegar, salt, sugar and pepper, and then bring to a boil. Once the sugar has dissolved and the brine has come to a boil, remove from the heat and let it cool slightly. Pour the brine into your prepared jars, making sure to cover the vegetables completely. Seal and place in a boiling water bath for five minutes (no longer, or pickles will be soft.)

Step Three: Let the Flavors Blend

Once the jars are cool, let the brine pickle the vegetables for at least four days before you eat them. Once they are brined, they are safe to keep in the refrigerator for up to one month after the jar has been opened.

Step Four: Eat Up!

These homemade pickles make great gifts and are an excellent addition to any sandwich or cheese platter. Enjoy your homemade pickles!

Pickles made with vinegar are delicious, but there is a third way to make pickles that contributes a very important benefit: probiotics. Making naturally fermented pickles at home may sound daunting, but if you are interested in fermenting for its many health benefits, you will be happy to know that pickling is one of the easiest and most delicious ways to add healthy probiotics to your diet. All it requires are a few fresh cucumbers, a clean glass jar, salt, and spices. For complete instructions, follow the link in Figure QR 1.5.

When you bite into your first homemade, fermented pickle, you will be pleasantly surprised by its crunch and flavor, so different from a store-bought or vinegar varieties. Subtle flavors of garlic, dill, or your preferred mix of spices combined with salty zip and mild sourness (created by lacto-bacilli) will delight your mouth and your gut.

FREEZING

One of the easiest ways to preserve foods is to store them in the freezer. Some produce can be frozen raw, while some should be blanched (lightly boiled) first. Some fruits freeze well in a simple sugar syrup, like raspberries, plums and cherries. Check online the first time you freeze something new. An internet search, e.g. "how to freeze raspberries", yields great results.

Remove as much water as possible before freezing. A salad spinner comes in handy for drying produce, or dab them with paper towels. Wrap meats in freezer paper and seal with freezer tape. Vegetables and fruits can be stored in freezer bags or boxes. Vacuum-sealed bags store the longest with the least amount of flavor and nutrition loss.

Always label the produce you freeze, including the type of food and the date. Months later when you're sorting through a full freezer trying to find those berries, you will be glad for the labels! I also try to stack similar items together in the freezer so that I can find what I want easily.

Frozen foods keep for a long time, but not forever. Most will start to lose flavor after eight to twelve months. When you thaw frozen food, be safe. Thaw in the fridge overnight or *quick thaw* in a warm water bath.

Here are a few of my favorite foods to freeze, and some tried-and-true methods I have used:

SWEET CORN: Blanch fresh cobs in boiling water for four minutes. The sooner that you blanche them after harvesting, the better and the sweeter the end result. Cool the cobs quickly in ice water. Cut the raw kernels off the cob in a large bowl. Sprinkle them with a little sugar and a little salt, and quickly freeze in bags or boxes.

BERRIES: Berries are so easy to freeze. Just remove stems, wash and dry thoroughly, and spread them out on a cookie sheet lined with wax paper or parchment. Freeze overnight. When the berries are completely frozen, transfer them quickly to freezer

bags or boxes, and return them to the freezer. This method will achieve frozen berries that don't stick together in a clump. The technique works for peach slices, too. To prevent browning, dip peaches in a mixture of water and 10% lemon juice, dab them dry with a towel and immediately freeze.

PIE FILLING: If you're making a pie this summer, triple or quadruple the recipe and freeze the excess. One quart of filling is plenty for a deep-dish pie. Thaw the filling, pour it into your crust, and bake as usual. Hint: Combined with a shelf-stable graham cracker crust, canned pie filling makes a great gift, especially during the busy holiday season.

TOMATOES: Tomatoes can be frozen whole. When they thaw, they are mushy, perfect for making sauce, and the skins slide off easily. You can also make homemade tomato juice or spaghetti sauce to freeze. Another delicious option is to roast the tomatoes. Blanch tomatoes in boiling water for one minute, and quickly cool in ice water. Remove the skins, which should slide off easily after blanching. Cut tomatoes in half. Place them cut-side up on a baking sheet lined with parchment paper. Drizzle with olive oil and top with chopped garlic. Cook at 200°F for several hours, until tomatoes collapse and begin to dehydrate. Cool and pack in freezer bags or boxes.

HERBS & FREEZER PESTO: Herbs freeze well in individual portions packed in olive oil or water in ice cube trays. Freeze overnight, and transfer cubes quickly to freezer bags. Single servings of pesto are also a wonderful item to have in the freezer. You can make a large batch at once, and have pesto available for months. Gather the following:

- 2 cups fresh sweet basil leaves
- ½ cup olive oil
- ⅓ cup nuts (pine, cashew, macadamia, or your favorite nut)
- 2 cloves peeled garlic
- 1 teaspoon sea salt
- ¾ cup fresh grated Parmesan cheese

Mix all ingredients in a blender or food processor at high speed. You will need to occasionally stop the blender to scrape the ingredients down toward the bottom with a rubber spatula. Freeze in ice cube trays or small containers.

DEHYDRATING

Dehydrating is a great way to store produce that requires little space and no refrigeration. Drying foods can be accomplished in a cool oven or dehydrator, or sometimes on your kitchen counter when atmospheric humidity is low. Dehydrated herbs, fruits and vegetables should be stored in an airtight container in a dark place. My preferred method is to store dehydrated items in glass canning jars in my pantry. Some of our favorite foods to dehydrate are:

HERBS: Herbs dry quickly and easily in paper lunch bags. I will often harvest whole stems from my herb plants, tying the ends together and turning them upside down in the bags. I close the bags to keep dust and pests out, and place them in a sunny window. In a few weeks, I crunch the exterior of the bags to knock the dried herbs off into the bottom, remove the bare stems, and fold the bags into small herb packets. I use these herbs to refill glass herb shakers.

APPLE RINGS: Slice apples very thin. I use a mandolin slicer or the slicing blade that came with my food processor. Sprinkle on a bit of cinnamon and sugar, and spread out in a single layer in a dehydrator. Be sure to spray the dehydrator rack with oil beforehand, or the slices will stick! You can use wax paper or a Silpat mat to prevent sticking, as well. Apple rings never last long at my house, and they often get eaten in one sitting. However, they do store well in glass jars. But be warned – if they are not completely dry, they may lose their crispness in storage.

TOMATOES & HOT PEPPERS: Dehydrated peppers and tomatoes are great additions to sauces and other dishes. Slice in half and remove the seeds. Place in a single layer in the dehydrator or 200°F oven, cut side up, until nearly all the moisture is removed. To rehydrate, soak in water or simply throw them into your sauce while it is cooking. To make hot pepper flakes, freeze dehydrated peppers overnight. Frozen peppers are brittle and easy to crush for pizza or recipes.

KALE: Dehydrated kale makes a delicious substitute for potato chips. Tear kale leaves into 2in pieces and spread in a single layer in the dehydrator. Brush or spritz with olive oil and sprinkle with garlic, salt or your favorite spice. Process until the leaves are crisp. You can also dehydrate nutrition-packed kale and other leafy greens, crushing the finished product into shards or a powder. Use the powder to boost the vitamin and mineral content of soups, smoothies, and sauces.

FRUIT SNACKS AND FRUIT LEATHER: When my kids were small, they loved fruit snacks and fruit leather, but many commercial brands of these snacks contain a load of preservatives, artificial colors, artificial flavors, added sugar, and little, if any real fruit. Few organic fruit leather or fruit snack options exist, and those that can be found on the store shelves are pricey. Thankfully, fruit leather and dried fruit are easy and inexpensive to make, with or without a dehydrating machine. The drying process preserves fruit nutrients and concentrates the flavor of the fruit. Dried fruit stores for a very long time, and homemade fruit leather is extremely easy to roll up to take with you for camping, hiking or lunches. Dehydrating fruit is also an economical way to preserve surplus fruit when it is in season. The link in Figure QR 1.6 will take you to a few of my favorite recipes.

IN SEARCH OF *PLENTY*

Preserving foods is a step in the right direction towards great self-sufficiency and true convenience, but I would be remiss if I did not also encourage you to use what you produce to its maximum potential in two important ways. The first is to eat more than one part of the plants that you grow.

In my gardening endeavors, I have often felt the pain of hard work and expense for little reward. I am reminded of my initial attempts to grow broccoli, how excited I was in the beginning, only to be let down when I realized how much work had gone into producing one single head per plant. Relative to how much precious garden space and how many resources the plants needed to grow, such a limited harvest seemed a meager return. The problem was not the broccoli plant, but my view of its usefulness, which was limited to the florets. But I discovered that all parts of the plant are edible. When harvested young and steamed, the stems have a lovely texture and sweet flavor. I learned to slice the stems into medallions, which my children referred to as *broccoli stars*. The edible leaves are large and sturdy, useful for wraps. They are delicious sautéed, similar in flavor to cabbage. By expanding my perspective of what was good to eat, each broccoli plant yields numerous meals rather than only

one. Once the broccoli heads have been harvest, I leave the plants in the ground to produce side shoots and leaves for the rest of the season. When it is finally time to pull the plants and grow something else, they are chopped and simmered with other vegetable ends to make stock. Nothing goes to waste.

The second way to maximize the potential of our garden produce is to use plants to make things that would normally be purchased commercially. I believe that our current food distribution system has played a role in blinding us to the bounty that is all around us, and it essentially trains us to go to the store and spend money for everything that we need.

The best way that I can explain how this works is with a specific example from my own life. I have a grapefruit tree that produces lovely red fruits. We harvest baskets of fruit off of the tree, so much that we give them away by the dozens. When the tree was young and I was new to growing citrus, we always seemed to have fruit that would go to waste. And I was never certain that the tree's water and fertilizer consumption outweighed the amount of food that we were actually getting from it. We often resorted to juicing the fruits, and it bothered me that pounds of pulp, rinds, stems and seeds were discarded in order to produce a relatively small amount of usable liquid.

I cruised the internet to find ways to use all of this waste, and what I discovered revolutionized my thinking. One of my first grapefruit exploits was to turn the rinds into marmalade. In my search for recipes, I happened upon an article that explained how to make pectin out of citrus seeds and the bitter white pith that grows between the skin and the fruit. Pectin is a thickener that I use often for making preserves, and I always kept a stock of commercial pectin packets on hand. It dawned on me that I had been essentially conditioned to purchase pectin at the supermarket. In fact, before I read the article, I had no idea what was actually in a packet of pectin or how it was generated. And it never crossed my mind that I could actually *make* it at home.

The idea that I could be more self-sufficient, making something that I needed at home using seeds and pith that would normally end up in my compost pile, was an epiphany. Now I had a nutritional use for every part of the fruit.

Over the years, we have made dozens of batches of grapefruit rind candy, grapefruit curd and marmalades. We have also soaked the rinds in vodka to make a martini base, and have dehydrated and crushed the pith for citrus powder that we add to our smoothies. Leaves are dried and stored in jars for tea and Thai cooking. Rinds are soaked in vinegar to make a citrus-y household cleaner and grated to make a natural sugar body scrub. So even when the fruit is out of season, we have grapefruit-based products on the shelf all year long, and my trips to the store are diminished. Figure 1.5 displays a few of the grapefruit products produced at the farm.

1.5

GRAPEFRUIT

DELICIOUS EDIBLE PERENNIAL

ALSO GOOD FOR MAKING MARMALADE, CURD, CITRUS POWDER, CANDIED GRAPEFRUIT RINDS, ALL-PURPOSE CITRUS CLEANER, CITRUS SUGAR SCRUB, ETC

With each successful attempt to try something new, the desire to use every part of my garden plants and to make more of the products that I previously purchased at the store has grown. When I plant something new, I enjoy researching and experimenting to find more uses for the plant than are typical. To my delight, I have discovered that citrus wood is wonderful for smoking meats, that Swiss chard stems make delightful relish, that apple cores are the main ingredient in homemade apple cider vinegar, and that even the cochineal scale pest that attacks my cacti are useful for making an exquisite magenta dye.

I believe that if we will enlarge our mindsets and learn to recognize the plenitude that is all around us, we can grasp hold of true abundance. And the concepts of scarcity and lack can virtually disappear from our food stores and from our thinking.

STRATEGIES FOR LIVESTOCK AND POULTRY

Backyard livestock adds a whole new dimension to urban farming, expanding the options for food, fiber and other animal products. Fresh eggs, raw milk, cheese, meats, wool, feathers and manure fertilizer are just the beginning of the list of benefits that animals can provide. Though greatly rewarding, livestock also demands a great deal more responsibility than garden plants require. The daily tasks of caring for animals can be fun, fulfilling and fruitful. They can also become a drudgery and a costly endeavor. What makes the difference? Just as it is in gardening, the door to prosperity swings on the hinges of prior planning and preparation. Here is what to do before you bring animals home to the farm.

- Gain knowledge and contacts. Learn as much as you can about the animals that you want on your farm before you acquire them. Take classes, read publications and talk to people who have experience raising urban livestock. Volunteering to help out with a 4-H or Future Farmers of America group is a great way to gather information and experience. Your local university extension office may also be helpful. Knowing what supplies you will need, the best places to acquire feed and supplies, and the rules for keeping animals in your city will lay the essential groundwork for hassle-free and economical farming. Having someone to call upon as questions or emergencies arise and the phone number of a reliable livestock veterinarian will give you peace of mind should an animal become injured or ill. And should you want to take a vacation, or even need to leave suddenly, having a list of dependable people who can tend to your animals will save you the trouble of finding a caretaker at the last minute. The more knowledge and relationships that you build in advance, the better prepared you will be to meet the challenges and reap the benefits of raising livestock.
- Prepare enclosures and pens. Strong fences are paramount. Make certain that your animal housing is secure from weather, as well as sturdy enough and tall enough to keep predators outside and livestock inside. Predators easily breach chicken wire and break through the smallest of gaps in a fence. Goats are notorious for climbing chain-link and finding the weak spots in enclosures, and they spend all day long looking for a way of escape. Our goats would also rush the gate as I opened it, bulldozing by me to rush out into the yard. We prevented that situation by adding a safety catch, which is essentially a small pen situated outside the main entrance to the animal enclosure. The catch served to block them from escape, should they overpower me at the first gate.

- Dependable, animal-proof latches are equally as important. Our ram, Oliver, spent a lot of time tinkering with the bolt lock on his pen. Much to our surprise, he was eventually successful in opening it, despite a secondary latch that had to be lifted before the bolt was able to slide. Our dog also figured out how to open a bolt lock on our chicken coop, and she used this skill to steal eggs from the nesting boxes. Never underestimate the ability of livestock or predators to open catches. By contrast, locks should be as effortless as possible for humans to use. Nothing is more annoying than a latch that is difficult to operate, so choose them carefully. Keep them in good condition so that they readily slide, turn or catch as intended.
- Explore water delivery systems. Easy-to-fill and easy-to-maintain equipment ensures that animals have dependable access to fresh, clean water without a lot of effort on the part of the farmer. When we started our urban farm, we purchased inexpensive equipment to keep expenses down, such as plastic chicken fountains that were a hassle to fill and were prone to algae growth. It may not sound like a big deal, but my water source was located 25ft outside the chicken run, so we either had to carry the fountains to the bib or drag a hose into the run to clean and fill them. The fountain jugs were too heavy for my kids to handle, so the job fell primarily to me. During the summertime, I was going through this process twice daily, and it became a grind. To streamline the operation, we hired a plumber to install a hose bib adjacent to the run. I now have an automatic watering device connected to the spigot, as well as a dedicated hose to fill up the birds' backup water dishes and to water the tree that is growing inside the run. I

1.6

COOP *OOPS*

DON'T GO CHEEP!
Pun Intended.

HINGES THAT WERE TOO SMALL FOR THE WEIGHT OF THIS DOOR RESULTED IN A MINOR DISASTER...AND MORE WORK THAN WOULD OTHERWISE HAVE BEEN NECESSARY

highly recommend using auto-watering equipment whenever possible, especially equipment that is easy to clean and maintain. And, in my opinion, it is worth the expense to have spigots installed adjacent to livestock pens as this will streamline daily watering and cleaning tasks.

Experienced contacts, education, sturdy pens and easy watering equipment will put your mind at ease and allow for farming that is as hassle-free as possible. This will make farming more fun and more sustainable, preventing burn-out. You will be a better informed and better prepared caretaker, able to handle health issues and other crises as they arise with the help of the network of contacts that you have put in place. Your animals will be safer and more secure, allowing you to farm with confidence, joy and peace of mind. And you can avoid the situation that we encountered in Figure 1.6.

PLANNED EXPANSION

It is said that chickens are the *gateway animal* to other types of livestock, and this was true in our case. Once I had acquired hens, I had the fever to bring more animals to the farm. On more than one occasion, I put the cart before the horse and brought livestock home before we had housing, or even a plan for where the housing would be located in the yard. This was usually due to my optimistic spontaneity, so when I made a spur-of-the-moment purchase of five turkey poults, I had full confidence that we could figure out the housing situation. I knew that my good friend Christy often biked in an area where bulk trash was scheduled for pickup, at which time people pile up all sorts of interesting items on the curb, waiting for the city trucks to haul them away. She offered to keep an eye on the piles to spot useful housing materials for me. As luck would have it, she found a sturdy blue plastic kiddie pool that same day, and we picked it up for free. The pool was situated inside our shed, wrapped in chicken wire, and filled with straw. Figure 1.7 shows the pool and two of my girls with adorable poults. This served as the poults' home until they were big enough to escape the pen, which happened with much greater speed than I had anticipated. We moved them into a 5ft by 5ft rabbit hutch, which they again outgrew with lightning speed. To accommodate their rapidly increasing size, we made a large fenced enclosure in the yard out of 5ft tall hardware cloth with covered roosting bars at one end. The pen was ugly, and the turkeys would often escape it. Nevertheless, it remained their home for two years.

1.7

EXPANSION

A LITTLE PLANNING GOES A LONG WAY

WE COULD HAVE RAISED OUR TURKEYS IN A MUCH BETTER BROODER HAD I NOT BROUGHT THEM HOME ON THE FLY

Much to the chagrin of my husband, Lewis, and his buddy Tim, who were called upon to assemble many structures on the fly, quail, goats and sheep would appear quite suddenly and quite often on the property. All kinds of used materials were repurposed for makeshift livestock housing, sometimes in a panic. On one occasion, it occurred to me that our young ewe and ram were old enough to require separate housing ASAP. We hastily assembled two dog kennels for them, and since it rains so seldom in Arizona we postponed creating roofs over the structures. Naturally, it rained the following Monday when Lewis was at work and the kids were at school. Left to my own devices, I threw on my slicker and ran out into the rain to find materials with which to create coverings. Within an hour, the pens were hooded with tarps and discarded political signs, which were secured with bungee cords and zip ties. Though it was unsightly, it kept the sheep dry for the moment. And although the makeshift roofs were only intended as a swift and temporary housing measure, they remained as they were for two years.

I don't regret the haphazard manner in which our farm was created; it was quite an adventure! But if I had it to do all over again, I would take the time to plan and prepare much better. Fortunately, I have had the pleasure of using my experience to help several friends in designing and constructing lovely farms that meet the needs of their livestock while also being attractive and inviting. With some planning, housing structures can be inexpensively constructed, built to last and easily maintained.

HARVEST PLANNING

Planning for the harvest is the key to efficiently using all of the bounty of a garden, and it is just as important to livestock production. Milk, eggs and meat arrive seasonally, sometimes yielding overabundance that must be harvested, prepared and stored in a timely manner. Taking the principles that we discussed in relation to garden harvest planning, the practice starts with knowing how you intend to use the harvest. Gather fresh recipes as well as preservation techniques and equipment in advance.

Mark your calendar to determine when to expect or when to schedule the harvest. Some animal products don't require very specific harvest date planning. Eggs may be in short supply in the winter, but overabundant during the summertime. Milk is available for months or years following kidding. However, meat animals are harvested at specific ages with a short window of opportunity. Whether the growth period is five weeks for meat chickens or ten months for sheep, the time will pass quickly and the harvest date will be upon you before you know it.

Prepare all your tools and supplies in advance to avoid a last-minute scramble. Learn how to do the processing by attending a class or assisting someone else. And if you will need help, have your collaborators lined up beforehand. I highly recommend having someone on hand for emergencies, as well. I have heard more than one instance in which meat birds started dying in a heat wave, requiring emergency processing before they all expired. Although I have not had that particular experience, our ewe prolapsed beyond repair quite suddenly. We needed to process her quickly, both to preserve the meat as well as to put her out of misery. We were not expecting to process her ourselves; neither did we have the equipment or the skills to do so. Very luckily for us, we were able to contact an acquaintance from our local feed store who was willing to give us a hand. Although the harvest, performed in a moment of crisis, did not result in precise, professionally packaged butcher cuts, we had meat in the freezer that would otherwise have gone completely to waste.

PRESERVATION METHODS

RAW MILK: If you are overwhelmed by copious amounts of milk, you can freeze it for later use. Glass jars are the best storage containers for milk, but they tend to crack if not handled properly. To freeze safely, pour chilled milk into sterile glass canning jars. Fill each jar only two-thirds of the way to the top, as milk expands greatly during the freezing process. I recommend using wide-mouth jars, as they have less tendency to crack under the pressure of the swelling liquid. Place jars in the freezer, leaving the lids off until thoroughly frozen. To thaw, place in the refrigerator overnight. Raw milk can be stored in the freezer for several months, although small texture changes are to be expected. The longer it is frozen, the grainier the milk becomes. To restore some of its smoothness, shake jars vigorously as they thaw. Milk that has been frozen is suitable for cooking, baking and cheese-making.

QR 1.7

CHEESE: The milk from our Nigerian Mini Goats is sweet and creamy, with none of the musky flavor associated with store-bought goat milk. I love to drink it raw, just as soon as it is chilled. But when our three does are in season, we have more milk than we can possibly drink before it spoils. This is by no means a problem; we love having the excess for making yogurt, sour cream and cheese. One of my favorites is Feta, which you can learn to make by following the link in Figure QR 1.7.

When I first attempted making cheese, I was a bit intimidated. But I quickly learned that anyone who can follow a recipe can make cheese. For many cheeses, the only equipment you need is a stainless steel pot, measuring spoons, cheesecloth and a dairy thermometer. You may also need cheese cultures and starters, which include bacteria, molds and acids that coagulate milk and give different cheeses their unique flavors. Cultures are readily available from online specialty stores, as well as Amazon and eBay. Figure 1.8 shows a wedge of bleu cheese made at the farm using a culture.

1.8

Homemade Bleu Cheese

Cheese-making recipes vary, from quick and easy to more challenging. Start with a simple one, such as yogurt, which is basically a very lightly fermented form of cheese. Easy instructions are posted at the link in `Figure QR 1.8`. Give it a try! Then graduate to cream cheese or Ricotta. Mozzarella or Feta require a few more steps, but are also pretty easy to make. Recipes abound online.

THE DIRT
on farming

Yogurt Recipe

CCITYFARMINGBOOK.COM/HOMEMADE-YOGURT-RECIPE

QR 1.8

`RAW EGGS:` We keep somewhere in the neighborhood of twenty-five hens at The Micro Farm Project. During the springtime, these hard-working girls can produce two dozen eggs daily, which is more than our six-person family can possibly eat. What we don't sell, I freeze for the winter months when egg production decreases markedly. To freeze eggs, whisk a dozen to thoroughly mix the yolks and whites. Do not whip to a froth as the air bubbles are undesirable for freezing. Twelve eggs should yield approximately three cups of liquid. To prevent the eggs from becoming grainy when frozen, stir in 3 tbsp of sugar *or* 1.5 tsp of salt. Fill each compartment of an ice cube tray or lined cupcake pan two-thirds full of the egg mixture. Freeze until solid. Remove frozen cubes, and package in moisture resistant containers, such as freezer bags. Seal and return to the freezer immediately. One cube from the ice cube tray yields one egg, and one cube from the cupcake pan yields two eggs. Use as you would fresh eggs for baking or breakfast.

`BONE BROTH:` The tastiest liquid you will ever sip from a mug, bone broth provides all the comforts of a hot, steamy cup of broth. Unfortunately, the convenience of canned and cubed bouillon has caused us to forget the value and flavor of slow-cooked broth that has simmered long enough to extract the *magic* from the bones, an art that our predecessors understood and practiced by necessity before commercial broths became available. We settle, instead, for relatively tasteless broths, devoid of nutrients, and sometimes full of additives and MSG. But when we have bones, meat scraps, and vegetable ends, making broth is so easy, creating a delicious concoction that is healthy for our bodies, uses up scrap materials that would otherwise go to waste, and provides a kitchen staple that stores easily for long periods of time. Follow the link in `Figure QR 1.9` to make your own broth. In my experience,

QR 1.9

having broth on hand has averted many trips to the supermarket, and I always keep some available. To freeze broth, let it cool and follow the instructions for freezing milk. Broth can also be canned using a pressure canner. Heat broth to boiling and pour it into sterile glass quart jars, leaving an inch of headspace. Process in the canner at 11lb of pressure for twenty-five minutes. Either frozen or canned, preserved broth will last for many months, supplying a healthy and quick base for making broths, soups, stews and gravies.

GETTING MORE FROM LIVESTOCK

Producing livestock taught me the high value of meat and animal products. They require commitment, effort and ample resources, not to mention the emotional element attached to raising and processing animals. We have made an effort to only eat animal products that were produced on our own farm or purchased from sustainable local farms. We keenly feel the cost of *clean* meats. Consequently, we eat less of it than we did prior to our farming experience, and we try to get the most out of our animals.

Recalling the manner that I research ways to get more out of the vegetables that we grow, I do so to an even greater degree with our livestock due to the heavier commitment involved with raising them. Fortunately, the greater commitment occasions even greater rewards when I am attentive to the possibilities. Besides providing meat, eggs and milk, animals can contribute by-products that support the ecological health of our property. Whenever I rake a goat pen, I am not just cleaning; I am harvesting compostable resources. When I crack an egg, I am not just making an omelet; I am collecting minerals for my tomatoes. And when I pluck a turkey, I am not just dressing a bird; I am gathering a slow-release nitrogen source for the garden. The City of Phoenix considers all of these by-products to be waste, and would prefer that I bag and throw them away. My opinion is at variance with city policy, not viewing them as trash, but as valuable raw materials that support the health and sustainability of our farm.

As you consider the options for your farm, think through and research all the possibilities that various animals represent. Perhaps you want poultry for eggs, but what more do livestock birds offer? Goats and sheep produce delicious raw milk, but what else can they make or do for the farm? Rabbits and pigs are superb meat sources, but are they useful in other ways? We will explore these questions in later chapters, and you can begin to do some of your own research on the web or by consulting with other farmers who raise the types of livestock that interest you.

INVENTING YOUR FARM

In your garden notebook, write down four or five foods that you want to produce on your farm. Leave some room between each item for notes. Use the internet, cookbooks or the advice of friends to gather a few recipes for cooking the items fresh and a few ways to preserve the excess. Paperclip the recipes into your notebook, or write down links, page numbers and other information that will allow you to access the recipes easily when you are ready to use them.

Include in your notes the equipment that you will need, as well as additional supplies and ingredients. Whether you will borrow or buy the items, begin to gather them in advance.

Finally, don't forget to consider how you will use the whole plant or animal. Probe the possibilities to determine what you will do with the roots, stems, and other parts that are typically discarded, but that may be edible or put to use elsewhere on your farm.

Jotting this information down now will serve to motivate and inspire your urban farming endeavors, and will give you a clear target at which to shoot. It will also serve to streamline the process of using all of the harvest, making it easier to capture the bounty that you will produce.

CONCLUSION

Growing and raising food in the city takes knowledge, creativity, and most of all, a commitment to provide the healthiest foods possible for ourselves and for our families. *City Farming* is all about helping you to acquire the skills and techniques to create superabundance on your own property. But if the resulting bounty is not used in a timely, purposeful and thorough manner, the effort is wasted. Conversely, applying the same levels of diligence and ingenuity to *using* the harvest as to *producing* the harvest materializes the very dreams that incited the desire to become an urban farmer in the first place.

Now that you have a plan to capture the bounty, the next chapter will begin to explore the basic skills and techniques to produce an abundant harvest. We will start with the sun, answering such questions as how much sunlight vegetables need, where to place your garden in relation to the solar aspect, and what to grow in shaded areas.

FEATURED FARM
LONGEVITY GARDEN

Vegan and athlete are two terms that one generally doesn't hear together. Not so with Jake Mace, the Vegan Athlete and proprietor of Longevity Gardens. Jake is known for two things: martial arts expertise and gardening. He has a large following in Phoenix concerning both topics, as well as a formidable presence on YouTube. What impresses me the most about Jake is that he eats every day from his garden, and that a large portion of his diet comes from produce grown on his urban farm. Figure 1.9 shows one of Jake's beautiful fig trees that is beginning to fruit.

When asked about the secret to his gardening success, Jake responds humorously, "I was telling my wife the other day that I really wanted to be an Olympic athlete growing up. And so, if I'm not going to be in the Olympics,

1.9

EAT YOUR YARD

JAKE REMINDS US THAT IF WE GROW FOOD, OUR JOB IS TO EAT IT! OTHERWISE, IT'S JUST ORNAMENTAL.

I'm gonna at least make gardening my Olympic gold event … So I attack gardening kinda like an athlete, where I really go at it intensely. But my whole purpose, my whole goal that motivates me is I wanna grow my own food. I really want so that I never have to go to the grocery store again. I still go to the grocery store for rice, and for cereal, and for almond milk, and stuff like that. But I don't go to the grocery store for produce, because I have all of it I need here."

Visiting Jake's garden is truly like visiting the Garden of Eden. Both the front and backyards are lush with healthy trees bearing fruit and nuts. A pond and stream run through a portion of the yard, and the ground is covered with a thick layer of wood chips. Raised bed gardens edge the swimming pool and cover a corner of the property.

It's clear to see the benefits of gardening in Jake's yard, but I was curious about what got him started. "I wanted to be vegan, and I also wanted to eat a lot of produce. Plus, I was teaching martial arts for a living and so I was working out full workouts, four times a day … I'm getting skinny because I just can't keep myself fed enough, and my food bill is $1,500 a month. And the economy is crashing, I'm 28 years old, so I'm the epitome of the millennial generation. I'm trying to make it, and everybody's out of a job, and I just can't afford to eat. So really my goal was to invest a year of growing food at home in order to have free food later."

When I interviewed Jake, he had only been gardening for a couple of years, and did not consider himself to be an expert. But with some learning and a lot of effort invested in spreading mulch, cultivating gardens and planting trees, he has created an oasis that has now begun to reseed itself and grow food efficiently. "I'm finally starting to get food for free because the garden is reseeding itself every season. And so I'm basically laundering money here. I'm printing my own eggplants and they're free."

Read the full interview by following the link in Figure QR 1.10.

Chapter 2

Sunny Side Up!

Optimizing Sunlight and Temperature On Your Farm

"Failure is only the opportunity to begin again, this time more intelligently. There is no disgrace in honest failure; there is disgrace in fearing to fail."

Shaw, Arch Wilkinson (1927). *The Magazine of Business* 52, 182.

I strolled lightheartedly across the farm on my way to the garden, inspecting my *estate*, a ¼ acre city lot. The modest size did not diminish the pride and satisfaction that I felt as I surveyed the neat gardens, contented hens gibbering in the background, the soft wind fluttering in the brim of my floppy hat.

Gathering tomatoes in my apron, I reveled in a few moments of private self-congratulations at the take, which had been a long time coming. It was a plum plucked from the pie of persistence; persistence necessitated by disappointment, disappointment born of failure; failure induced by ignorance, not in the ugly connotation of the word, but simply that I did not know what I did not know on the first attempt to grow food.

How different this day was from three years prior when I had worn the same apron, but had no harvest with which to fill it. I had tilled up a desolate corner of the yard, and as I stooped to plant in the dusty earth, the gardens in my imagination sprang to life, flourishing, blossomy, balmy and bounteous.

My fondness for gardens had begun decades ago in childhood, and in my salad days, as it were. Often in memory, I rambled through rows of vines, neck high in my grandfather's Wisconsin garden, brushing the saw tooth leaves lightly with my fingers, gliding by in search of ripe, red tomatoes. As I child, I knew nothing of the science of how a garden lifts one's mood. Nor could I have explained with a fledgling vocabulary what it was how I understood the latent vigor of the soil that permeated my being through bare feet, cooling my legs and quickening my spirit – how I longed to breathe in concert with the rhythm of life incorruptible, to invite the sun to energize my mortal body as only it can in a garden, to walk with God in the cool of the day, basking in easy companionship with the cultivator of my soul. All this was just a feeling, unformed with words, nudging me to wake before the sun, as soon as I heard my grandfather stirring, in hopes that he and I would go together at first morning light to pick tomatoes.

It's true that though I have beautiful garden memories, I had not actually gardened at all. My grandfather had been the cultivator, and I had enjoyed the fruits of his labors. Stepping blindly into his shoes, I worked doggedly, sometimes feverishly to vivify his Midwest garden in my Southwest yard. Gloriously, the spring plantings took root and grew. Soon, the tomato plants would be flourishing, perhaps even a little overgrown, and the tender lettuces would be crisp and sweet. But as the weeks flowed towards summer, my enthusiasm buckled like a collapsed soufflé. Intensifying sunlight and

scorching winds melted my enthusiasm like iron in a blast furnace as I witnessed my tender plants losing their will to live in the desiccating heat. Surrender came in June. I languidly removed my harvester's apron, wiped the sweat out of my eyes and considered that, although I had looked and acted the part of the Cottage Gardener, like a performer in a play, my attempt to grow tomatoes like grandfather's had not gone according to script.

The gray, dusty ground and the sight of wilted and dead plants popping up like pulls in an old sweater gave me a feeling of betrayal. Summers long ago in Wisconsin had always seemed like a benefactor, inviting me outdoors, rewarding me with good and easy company and a veritable Regent's Banquet of garden produce. Phoenix had offered no such summer indulgence. Contrary to my expectations of a lavish and verdant season, June ordered up broiling hot temperatures with a side of squash bugs, and '*86*' the harvest.

Retreating indoors, I pretended for some months that my gardening experiment had reliably proven the impossibility of growing food in the low desert. The evidence was at my back door, a garden dry, though watered; barren, though seeded; dead, though nurtured. I had plowed my garden, and it had plowed me.

But then, when all seemed lost, horticulturally speaking, something shifted. It's funny how mere coincidence can turn out to be so profound. In 2009 and in 2010, my husband, Lewis, was deployed by the US Navy Reserve. Left to myself, my kids in school, my garden failed, I was bored and unsupervised. Retreating to my easy chair, I became fixated on a new and consuming pastime – surfing the World Wide Web. Delightfully, one could enter any search term and receive a smorgasbord of information on topics, no matter how obscure. By happy chance, I typed in the fortuitous words *Gardening in Phoenix* ...

I don't remember everything I saw or in what order I saw it online, but pictures and stories and groups related to *Gardening in Phoenix* began to dismantle my belief that it was impossible to do, and evidence that it could be done stoked the fire of desire in me once again. If others could grow food here in the arid low desert, I could, too.

My courage was bolstered to apply for the Master Gardener program offered by The University of Arizona Maricopa County Cooperative Extension. If you have ever wondered what the training is like to receive the prestigious title of Master Gardener, there is no glamour involved. Fifty-one hours of lectures, sixty hours of hands-on

gardening (of which I actually did eighty hours,) and an overwhelmingly long final exam earned me the moniker. As rigorous as it was, the course only scratched the surface of what there is to know about gardening. But like an expertly crafted infomercial, it hooked me and raised my hopes that what I desired was possible.

The following spring, as advised by my coursework, I placed my tomatoes in a warm sunny location and my lettuces in partial shade, planting them early, much earlier than I remembered planting in the Midwest. Through February and March, we harvested greens and trellised the tomato vines that stretched and grew. As the ferocious heat of June approached, I was ready to defend the precious, blossoming tomatoes with shade cloth, a product that we had never seemed to need in Wisconsin. It all seemed too simple. I was pleased with the lettuce harvest, but still I wondered if the expectation of tomatoes was another setup for disappointment. Hope was rewarded and we were almost glad when production slowed around the Fourth of July. That summer, I was transformed from gardening naysayer and skeptic to promoter and enthusiast.

Since those early days, I have often been asked to give gardening tips for interviews and articles, and I think back on my first training and the simple advice that I received concerning sunlight and shade in a desert garden. For me, it was an epiphany to realize that shaded areas of the garden could be as useful as the sunny locations recommended in general gardening advisement. This tidbit of knowledge seemed to open up the heavens for me, and my next fumbling steps in the garden were congratulated with a thrilling harvest.

So important is understanding the role of sunlight in the garden that I chose to put it first in our study of growing techniques. Whether your property is hot and sunny, cool and heavily shadowed, dappled with light shade, or a combination thereof, this chapter will reveal how you can use all of these locations to create a prolific garden that thrives in any almost area of your yard.

THE EFFECTS OF SUNLIGHT AND HEAT IN THE GARDEN

It's no surprise that the amount and intensity of sunlight play an important role in fruit and vegetable development. Most fruiting plants, including many common vegetable varieties, need to bathe in the sun for six to eight hours per day in order to maximize fruit production. But the label *Full Sun* on a seed packet can be misleading; many plants, such as herbs, beans and peas, may grow just as well in partial shade. And some veggies that need lots of sunlight, such as tomatoes, can be scorched by the western sun in climates where afternoon heat is intense. And some plants are fickle, having light exposure preferences that change over time. Leafing lettuces prefer partial shade and cool temperatures for head production, while the seeds require exposure to warm sunlight in order to germinate. In reverse, mature summer squash enjoys intense sunlight, while the seeds germinate deep in the soil, no light required.

To make matters more interesting, plants tend to have optimal temperature ranges for seed germination, flower set and fruiting. In order to grow these finicky plants, gardeners sometimes have to take extraordinary measures to provide the right conditions for their garden crops.

SEASONS BLEATINGS

Plants are not the only ones on the farm that are sensitive to changes in sunlight and temperature. Fluctuations in day length and weather have their effect on livestock, too. Poultry and quail produce the most eggs when days are long and temperatures are mild. Some species of goats and sheep are fertile only during specific seasons. And soaring temperatures relate to high mortality for rabbits.

THE BASICS

Day length, the intensity of sunlight, and temperatures vary greatly from climate to climate. This affects animals and plant varieties in numerous different ways. Like putting together the pieces of a jigsaw puzzle, the more plant varieties and livestock species that exist together on a farm, the more complicated and intricate the system becomes. So, what's a grower to do, particularly a new urban farmer with big dreams but relatively little experience?

First, understand the basic sunlight needs of plants, which is not very complicated. Plants need light for photosynthesis, the process by which plants convert solar energy into sustenance. A substance called *chlorophyll* that is found in plant leaves converts

light into food in the form of carbohydrates. In order to perform photosynthesis and achieve optimal production, vegetables need six to ten hours of sun exposure per day. Full sun is great. But plants can also thrive in environments that provide a combination of full sun, filtered sunlight shining through tree branches, or reflected sunlight. Fruiting plants, such as squashes and melons, tend to need more sunlight than plants that produce edible leaves, roots or bulbs. Figure 2.1 says that morning light is ideal, and this is true in part because it lacks the heat of the afternoon sun. But any sunlight will do, no matter what time of day.

2.1

SUNLIGHT

6-10 HOURS PER DAY

MORNING SUN IS IDEAL

FRUITING PLANTS NEED MORE SUN THAN ROOTS, BULBS & LEAFY GREENS

Plant labels will often give you a clue as to how much light a particular plant needs. The term *full sun* means that the plant needs and can tolerate more than six hours of sunlight per day. Plants that prefer *partial shade* or *partial sun* need less sunlight, between four and six hours each day. Plants that require full shade should only be exposed to four hours or fewer of direct sunlight.

Secondly, now that you know the basics of sunlight, you can use this information to begin planning the placement of your farm elements. The following are simple steps to follow that will help you to make good placement decisions.

INVENTING YOUR FARM

STEP 1: OBSERVE

Day length and the intensity of the sun's rays vary seasonally. Mid-summer, the days are long and sunlight can be intense, shining almost directly from overhead at noon. Mid-winter, daylight hours dwindle and the intensity of sunlight softens, shining from a low, southerly angle in the sky.

Take some time to walk your property at different times of day and in different seasons, and start to keep a journal of what you observe. Jot down temperatures, warm areas in your yard and cool areas, as well windy spots. Note shaded and sunny locations, and how many hours of direct sunlight each area receives daily throughout the year. It may be helpful to draw a simple map of your property to keep track of the data.

STEP 2: START SIMPLY

Decide what you want to grow, initially selecting just a few varieties. Think about what you really like to eat and what you buy often at the grocery store. My standard recommendation for new gardeners is to cultivate a few herbs, which are expensive to purchase fresh at the market. To herbs I would add garlic and onions, which are extremely convenient to have at your fingertips. Herbs, onions and garlic are highly adaptable to light conditions, as well, thriving in full sun or partial shade, so they are wonderful plants to grow while you are still observing and assessing the light conditions on your property.

STEP 3: GATHER INFORMATION

I highly recommend that you connect with nearby gardeners and farmers who can advise you about local growing conditions. Run your list of items that you want to grow by a seasoned gardener who can tell you which ones will be easiest to grow in your climate.

Use the internet or ask at your neighborhood garden center for information about community garden clubs and classes. A number of helpful resources are provided by University Extension Services that are offered in every state. These programs provide free and low-cost classes, clubs and hotlines for gardeners and farmers. One notable program, mentioned in the opening paragraphs of this chapter, is the Master Gardener Program. The course is offered through land grant universities in the United States and Canada. The program presents intensive home horticulture training to individuals who then volunteer to provide regionally specific advice on gardening

in their communities. You can find local Master Gardener programs in the United States and Canada by visiting the American Horticultural Society website online at www.ahs.org/gardening-resources/master-gardeners.

STEP 4: EASTWARD HOE!

Use the information that you have collected to assist in making placement decisions for your farm elements. In the Northern Hemisphere, optimum vegetable garden placement is typically east-facing or south-facing in an area that receives six to ten hours of sunlight daily. In very hot climates, add afternoon shade protection to the list of ideals for summertime gardens.

Livestock animals generally thrive in habitats that allow them access to both sunlight and shade throughout the day when the weather is hot, and lots of sunlight for warmth in the winter months. This is easily accomplished by placing coops and pens in areas that are shaded by deciduous trees, such as apples and stone fruit varieties. These trees are green during the summertime, providing respite from the heat. They lose their leaves in autumn, allowing maximum sunlight and warmth during the winter.

STEP 5: BEND THE RULES

If *ideal* locations do not exist on your property, *do not despair!* In reality, no location is ever ideal. Every property has its challenges, particularly in an urban setting. It is exciting for a desert gardener, like myself, to realize that vegetable gardens and livestock thrive in every corner of our planet, even in the most extreme environments and under the harshest conditions. That's good news considering the harsh climate where my farm is located. And even though my farm is situated in the low desert, not every corner is hot and dry. There are areas with lots of vegetation that are shaded and relatively cool. Areas that differ from the climate of the general region are called *microclimates*. On properties that have more than one microclimate, some may not fit the *ideal*. But they still have great value to an urban farmer!

The key is to change the image that we hold of what is *ideal*, and to recognize that every location on our properties has its benefits and its drawbacks. As creative and thoughtful gardeners, we can discover the benefits inherent in *less than ideal* locations, in order to maximize their potential.

The idea that 'every problem *has* a solution' is transformed by one, single, powerful word to become 'every problem *is* a solution'. Throughout the remainder of this chapter, we will look at various challenges that face urban farmers in relation to light and

temperature, and how growers are putting all corners of their properties into abundant production by turning their problem areas into assets.

STRATEGIES FOR SHADE AND COOL TEMPERATURES

THE SUN I NEVER HAD

The word *shady* brings to mind something that is disreputable or suspicious. As the word suggests, common garden wisdom dictates that shady areas of our properties are to be avoided for food production. Let's call them *shaded* locations, as we will discover that these darker, cooler spots offer some advantages that sunny, warm areas lack. And, properly managed, they can be just as productive.

SHADY CHARACTERS

The first strategy for employing cool, shaded areas on your property is to plant vegetables, fruits and greens that thrive despite the lower light and temperatures. When selecting vegetables for shade, think *heads* and *roots*, instead of *fruits* that require more energy, and thus more light. Cool season vegetables and greens, such as cauliflower, broccoli, lettuces, kale, carrots and radishes, don't mind cool, shaded conditions. Many herbs can thrive in low light, as well as onions, peas and green beans. Figure 2.2 shows a partly shaded garden at The Battery Urban Farm in New York City.

2.2

PART SHADE

COMBINATION OF SUN AND SHADE, OR DAPPLED SHADE THROUGH THE TREES

GROW ROOT VEGETABLES, BULBS, BEANS, PEAS, HERBS, SUCCULENTS & LEAFY GREENS

If you garden in a hot climate, use the cool, shaded areas on your property to extend the growing season for your cool season garden. In particular, greens will last longer into the warm season in a shaded area than they will in full sun. Learn more about shade-tolerant vegetables and growing a shaded garden by following the link in Figure QR 2.1.

THE DIRT on farming

Crops to Grow in the Shade

CITYFARMINGBOOK.COM/CROPS-TO-GROW-IN-SHADE/

QR 2.1

SELECTIVE PRUNING AND REDUCTION

The second strategy is to create a sunnier area by pruning back shade plants and trees. As an example, The Micro Farm Project has a Mesquite tree next to a raised bed garden. When temperatures are hot, I allow the tree to grow and provide some shade to the garden. However, the canopy can become very thick, blocking out too much light. By selectively removing a few branches, dappled sunlight shines through the leaves, providing enough sunlight for the summer veggies. In the cooler seasons, I prune it back more severely to allow more light into the garden when the days are shorter and the angle of the sun is lower in the sky.

GIVE IT SOME STRUCTURE

Structures can create a warmer environment for garden vegetables. Some gardeners opt to build or buy a greenhouse to conserve heat and protect plants from the outside world. They are especially useful for seed-starting activities, for protection from the elements and pests, and for isolating varieties to prevent cross-pollination. Plans to build greenhouses using glass, Plexiglas, or heavy duty plastic are prevalent on the web, and small prefabricated greenhouses can be purchased for a few hundred dollars. But not all gardeners have the need, the room or the resources to have a full greenhouse. Fortunately, there are other structures that can provide the same benefits without the corresponding cost or complexity.

A cold frame is one such structure that is easy to build and simple in design, consisting of a removable glass lid that admits light into a four-sided frame. The frame traps warmth, and is placed on the ground to protect seedlings and small plants without artificial heat. On our farm, we created a portable cold frame using an old window that we found in our alleyway as the lid. The lid is removed partially or completely during the daytime to allow airflow, and closed at night. It works wonderfully to warm the soil enough to germinate warm season plants early in the spring. In cold climates that experience hard frosts, a cold frame may provide enough protection to grow cool season plants, such as greens, all winter long.

MIMIC THE FOREST

Plants that thrive in low-light conditions have often originated in forested areas that characteristically have moist, rich soils and plenty of organic matter in the form of ground cover. Recreate these conditions in your shaded garden areas by spreading a thick layer of mulch on top of the soil in the form of wood chips, straw, hay, pine needles or other stable material. In combination with deep watering, the mulch will serve to maintain moisture and nutrient levels in the soil. We will learn more about watering methods and mulch in later chapters. Figure 2.3 shows a heavily mulched shaded garden at Roosevelt Growhouse in Phoenix.

2.3

SHADED GARDENS

MIMIC THE FOREST FLOOR

PLANTS IN SHADED AREAS NEED SOIL CONTAINING AMPLE WATER AND ORGANIC MATTER

THIS GARDEN GROWS IN A SUNKEN BED FILLED WITH WOOD CHIP MULCH

MAKE A SIMPLE HOOP HOUSE

Small gardens can be protected from cold by the use of plastic sheeting. In essence, one can create a mini hoop house or greenhouse by building a simple frame to support clear plastic sheets. At The Micro Farm Project, we use steel remesh lattice to create the frame. Remesh, also known as hardware cloth, can generally be found at hardware stores in the aisle containing fence supplies. Remesh sheets are sturdy, but require two people to work them into place. Remesh lattice is easier to bend, and one person can set the supports alone. Bend a remesh lattice over the garden to form an arch, repeating at 2 or 3ft intervals. Press the ends of the remesh down into the soil to stabilize it. Drape clear, heavy plastic sheeting over the remesh supports, tacking it to the ground with landscape staples, heavy rocks or bricks. When temperatures rise above freezing during the daytime, lift up the plastic to allow airflow. Figure 2.4 shows a sheet of remesh arched over a raised bed at our farm.

WEATHER PROTECTION

PERMANENT SUPPORTS MAKE GARDEN PROTECTION EASY

PERMANENT SUPPORTS CAN HOLD FROST CLOTH, PLASTIC, OR SHADE CLOTH.

THEY ALSO ACT AS TRELLISES FOR CLIMBING PLANTS

2.4

As an alternative to plastic sheeting in milder climates, old bed sheets can be draped over the supports. Make certain that the plastic or the sheets are draped all the way to the ground when a frost or freeze is predicted to trap heat and prevent cold air from moving into the frame. Sheets should be lifted or removed when temperatures rise above freezing to release moisture buildup and allow for adequate airflow.

Remesh supports are an inexpensive way to create a protective frame for your garden that serves multiple purposes. Not only do the frames make frost protection easy, but the frames can also be used to support shade cloth in the summer and protection from hail, heavy rains and wind during periods of intense weather. I leave the frames in place all year long, and they make great trellises for climbing plants when they are not in use supporting garden fabrics. Follow the link in Figure QR 2.2 to learn more about easy weather protection methods.

A LITTLE REFLECTION GOES A LONG WAY

Make use of areas that lack direct sunlight by increasing the amount of indirect light. Reflection placed behind plants focuses and redirects light that is passing through the area back towards the center of the garden. To create reflection, paint the wall behind the garden white or hang a mirror strategically facing garden plants.

Reflective mulches placed on the soil are another tool for shaded garden areas. Red or silver mulches spread over the soil reflect light upwards towards plant leaves, increasing photosynthesis and plant production. Reflective mulches can be purchased online or at garden centers. A simple and inexpensive alternative is to cover cardboard squares with aluminum foil or Mylar, placing them on the ground around plants that need more light. Mylar sheets can be purchased online at any store that sells food storage supplies, or you can repurpose a Mylar balloon for use in the garden.

THINKING STRATEGICALLY ABOUT COLD SPOTS IN THE YARD

You may have an area on your property that is cooler than other areas. Cold spots often occur in areas that are shaded, that are low compared to the rest of the property, or that are exposed to wind. At our farm, there is a corner on the north side of the property that is heavily shaded in the summertime and which receives virtually no direct sunlight during the winter. The soil in that area is always cool. Though it is difficult to grow anything in that area during the winter, it is a great place to grow greens early in the fall and late into the spring when other areas of my property are too warm. It is also a wonderful, protected area in which we could place barrels for rainwater collection or a birdbath. I have also considered placing rabbit hutches in this area, as bunnies do not like to be exposed to hot temperatures. Cool, shaded areas might not be ideal for growing heat-loving garden veggies, but if we think strategically about them, they can be wonderful assets on the farm.

STRATEGIES FOR SUNNY AREAS AND HOT TEMPERATURES

Sunlight and heat go together, and our Arizona farm has some hot spots that are challenging during the summertime. I often joke that we garden in a blast furnace, and that the Valley of the Sun feels more like the *surface of the sun* during the hottest months of the year. The combination of intense sunlight, long days and high ambient temperatures can cause humans, animals and gardens alike to wilt in the summer. But though the temperatures often rise above 110°F, we are able to grow many plants even in the most intense areas of our yard. And we can strategically create areas on our property that are cooler than the surrounding climate for those plants and animals that need some relief. Here are some natural and economical strategies to make hot climates more comfortable for plants, animals and people, too.

FUTURE'S SO BRIGHT, I GOTTA ADD SHADE

Natural shade is perhaps the best kind for the garden. To illustrate, consider a sweltering afternoon and the choice to get relief from the sun by standing under an awning or in a bank of shade trees. I think that it's just more pleasant under a tree. This is probably due to the ability of trees and other plants not only to shade out solar heat, but also to cool the surrounding area via the evaporation of water from leaves, or transpiration. They provide a balance of shade and sunlight, creating a dappled effect that allows sunshine into the garden while tempering it. Besides providing relief from heat, shade plants have two other notable benefits. They act as a filter for contaminants, which is especially useful for a garden that is located near a road or other polluted areas, and they serve as a windscreen.

Since natural shade is more pleasant for my plants and also for me when I am working in my garden during the summertime, I employ tall plants to protect smaller, more delicate plants.

I am always on the lookout for ways to make gardening easier and more productive. So, if I were going to create the ideal shade solution for my garden, it would have the following characteristics: easy management, automatic shade in the summer and sunshine in the winter, aesthetic qualities, and added value to the garden (other than shade.) If I thought long and hard, I could not come up with anything better than deciduous trees and bushes.

Once established, deciduous trees require very little management, automatically growing leaves to shade the garden during the warm months; losing their leaves in the cold months when the garden requires more light. They add beauty to my landscape, attract pollinators, and shelter the garden from severe weather. The leaves that fall in autumn are superior materials for composting, which contributes to the overall health of the garden. Many varieties produce fruits and nuts. And there are many other benefits to the garden and to the environment, too numerous to mention here.

Vining plants provide many of the same benefits as trees and shrubs, while also remaining much more adaptable and less permanent than trees and shrubs. Like trees, many of them are deciduous and edible, such as blackberries and grapes. Others are annual plants, growing only for a season, such as squashes, climbing beans and melons. They are easy to grow on a trellis on the western edge of a garden or to temper any particularly hot locations on your property.

To illustrate the use of vines, at The Micro Farm Project, a cement block wall surrounds the property. This type of wall serves as a heat sink, absorbing solar rays and collecting heat during the daytime, and releasing it slowly during the evening. From May through October, gardens placed near these walls can suffer from the heat that radiates from the cement. During the cooler months of the year, these gardens remain slightly warmer and benefit from the added warmth. In order to temper the heat of summer, we grow grapevines and annuals on trellises against the wall. These plants love the summer heat, and they provide shade for the wall, which keeps our gardens cooler. During the wintertime, the vines are either dormant or have been removed from the garden, allowing the wall to fully absorb heat and radiate it back to the gardens when it is beneficial.

In addition to vines, certain annual plants are taller than others, and they can serve as shade and windbreak to the lower-growing varieties.

Many varieties of okra, basil and wheat are tall enough to cast a protective shadow on their more delicate neighbors. If you have a large growing space, corn is one of the best plants to propagate for garden protection. Tall corn tolerates the heat of summer and it need lots of sunlight for good production. Corn also requires wind for pollination. For these reasons, rows of corn are an excellent choice for shade and wind protection, planted on the west side of the garden or in the direction of prevailing summer winds. When growing corn, keep in mind that you will need a minimum of forty stalks planted densely to achieve good pollination, and even more stalks will be necessary if you space them around your property to shade different areas. Although corn takes up a lot of garden space, it can be interplanted with vining beans that use the cornstalks as a trellis, and with low-growing plants in between the rows. Follow the link in Figure QR 2.3 to see photos and learn more about how we use natural shade at The Micro Farm Project.

QR 2.3

Another shade option is flowers. Flowering plants add beauty to a garden and attract pollinators. They can also be very useful for shade and wind protection. In many of our garden beds, you will see sunflowers, hollyhocks, Hibiscus sabdariffa and Jerusalem artichokes planted densely on the western edge. Figure 2.5 shows Hollyhocks growing on the west side of a lettuce bed in the front yard at our farm.

2.5

LIVING SHADE

TALL PLANTS & TREES PROVIDE LOVELY, EFFECTIVE SHADE

AT THE MICRO FARM PROJECT, HOLLYHOCKS GROW ON THE WEST SIDE OF THE LETTUCE BEDS TO PROTECT THEM FROM THE AFTERNOON SUN

INVENTING YOUR FARM

If I have piqued your interest in planting natural shade, here are some pointers. Since trees and bushes are relatively large and permanent, place them carefully in your landscape:

1. **Location:** Plant deciduous trees and bushes to the west of your garden to provide afternoon shade during the summertime and full sun in the winter. Also, consider planting trees and bushes to temper prevailing winds. Observe your landscape over the course of a year to discover from which direction the strongest winds come. If prevailing winds originate in the southeast, plant a windscreen to the southeast of the garden. If they come from the north, then plant a windscreen to the north of the garden. Avoid planting trees due south, as they will provide very little shade during the summertime when the sun is high, and will block sunlight during the wintertime when the sun is low in the sky. Even deciduous trees should not be planted to the south. Although they lose their leaves and cast a minimal shadow, it may be enough to hinder garden production during the winter when vegetables need as much sunlight as they can get.
2. **Variety selection:** In your observations, note the seasons in which wind speeds are at their highest. If winds are strongest during the summertime, either deciduous or evergreen varieties are appropriate windscreen options. However, if winds are strongest any other time of the year, evergreen varieties, such as citrus or pine, are the best choice.
3. **Height and foliage density:** Trees and bushes vary widely in height and foliage density. Low-growing or sparsely branched varieties may be planted very close to the garden. However, tall and densely foliated varieties will cast long, dark shadows that may shade out garden production. These should be placed further away from the garden to ensure that plants receive a minimum of six hours of sunlight daily (or at least four hours of sunlight for shade tolerant varieties).
4. **Shading your home:** The same principles that apply to using live shade for your garden also apply to shading your home. Strategically placed deciduous trees can naturally reduce temperatures in your home during the summertime, and allow sunlight and corresponding heat into the home during the wintertime.
5. **If you haven't already drawn an aerial view map of your property, do so now.** The drawing does not need to be professional in appearance or to scale, but it should show the basic layout of your yard, including all major existing structures and landscape features. Once the map is

drawn, use a pencil with an eraser to sketch in potential areas for trees and gardens. This simple exercise will help you to discover the best possible placement of foliage on your property.

On our farm, we have five raised bed gardens in the front yard on the east side of our home in a very hot spot. The area receives full sun until about 6pm during the summertime, which is a very long day for gardens to survive the heat and searing sunlight of Phoenix. I planted a Mesquite tree to protect this garden. Since this variety only loses about half its leaves in the wintertime, I strategically placed it on the north edge of the beds so that it would not shade out any of the sun that enters the garden at a low angle from the south during the wintertime. In the summertime when it is in full foliage, it does a wonderful job of providing dappled shade.

In our backyard at the southern end of the property, I have another area of raised bed gardens that gets very hot in the summertime. A loosely branching Palo Verde tree provides shade on the north side of this garden, and small fruit trees are planted directly to the west. I purposefully prune these fruit trees to grow no taller than 7ft, which serves to make harvesting easier and prevents the trees from casting long shadows that would shade out production of the gardens situated closest to them.

SHADE CLOTH AND STRUCTURES

In our discussion about frost protection using plastic sheeting, I described the remesh arches that we use in our garden beds as supports. These same arches can also be used to support shade cloth during the summer. I recommend covering tender plants with loosely woven burlap or with shade cloth that has a UV protection factor of 30% or lower. This will provide relief to sensitive varieties, but will not inhibit photosynthesis or shade out their productivity.

Unlike frost cloth, shade cloth does not need to be removed daily. It can be left on the supports for the entire summer season. Shade cloth can be stretched like a canopy over the garden, or draped to the ground to serve the dual purpose of deterring pests. Note that draping cloth to the ground will prevent bees from accessing plants that require them for pollination. When that is the case, the sides of the cloth can be lifted up temporarily at dawn and dusk to allow access when bees are most active. Figure 2.6 shows a simple shade structure in a garden plot at the Phoenix Renews Community Garden.

2.6 Plot at Phoenix Renews Community Garden

SHADE STRUCTURES

USE FOR SEASON EXTENSION & TO PROTECT TENDER PLANTS FROM HEAT OR SUNBURN

SHADE CLOTH, BURLAP, ROW COVERS, TREES, & TALL GARDEN PLANTS CAN PROVIDE SHADE

I typically use shade cloth for two plants in my garden: greens and tomatoes. When the days start to lengthen in the spring, cool season greens are shaded by draping burlap over a structure, hanging the cloth all the way to the ground to decrease the temperature underneath the cover. This practice extends the cool growing season by a few weeks. Lettuces, kale and spinach will last a bit longer into the warm season by placing them in the shade where soil and ambient temperatures are kept a few degrees cooler. I also shade tomatoes, draping the cloth to the ground. In most climates, shading tomatoes may be counterproductive. But in the low desert, it is a necessity. The slightly cooler temperatures prevent tomatoes from essentially cooking on the vine, resulting in a harvest of stewed tomatoes. It also ensures that the tomatoes will flower longer than they would in full sun, and that tender pollen is not deactivated by rising temperatures before the fruits begin to form. Although bees are often seen buzzing around tomatoes, it is okay to drape shade cloth to the ground because bees are not necessary for tomato pollination. Tomatoes are self-pollinating plants, but I help them along by giving them a shake during the cool hours of the morning when pollen is active. The vibrating action releases and distributes the pollen, and the cool temperatures provided by the shade cloth ensure that it is viable.

WEST SIDE STORY: PLANTS THAT LOVE HEAT AND SUNLIGHT

Solar radiation is most intense at noon when the sun is highest in the sky. As the sun travels towards the west in the afternoon, the strength of its rays begins to soften, but temperatures can continue to rise for several more hours, peaking between 3 and 4.30pm. During the summertime, western exposures can become hot enough to wilt or even burn garden plants. If you notice that plants droop in the afternoon, perking up as the sun begins to set, or if brown, sunburned spots appear on fruits or leaves, it may be necessary to add a shade structure to provide them with some protection from the western sun.

Areas that are hot and sunny are not always problematic. In fact, they can be assets. Many plants flourish in full sunlight and high temperatures. Corn, beans, melons, squash, sunflowers, basil and oregano are a few examples. These are the varieties that I plant in areas of the farm that are unprotected from the sun. Figure 2.7 shows a garden in full sun, verdant even in the desert summer at The Sweet Life Garden in Phoenix.

2.7

FULL SUN

THE SUNNIEST LOCATIONS ARE OFTEN ON THE SOUTH SIDE OF YOUR HOME, & THE HOTTEST SPOTS ARE OFTEN ON THE WEST SIDE.

GROW SQUASH, MELONS, BASIL, OKRA, SUNFLOWERS, HOLLYHOCKS, CORN, GRAPES AND FRUIT TREES

Hot spots in the garden are also wonderful for season extension. The soil warms in these areas sooner in the springtime than in more shaded locations, resulting in earlier seed germination. And plants that thrive in the heat will last longer into the fall when grown in full sun in a warm location.

Previously, I mentioned that one of the hottest areas of our farm is adjacent to the cement block fence. During the summertime, I grow vines on the wall to abate the heat. But when temperatures begin to cool down and the vines go dormant, the wall serves as a heat sink that raises the temperature of the nearby area to extend the life of my warm season plants. Due to our mild winters, I am often able to grow tomatoes through the winter next to a west-facing wall with minimal frost protection. For me, fresh tomatoes harvested in the winter are the consolation prize for living in a desert.

STRATEGIES FOR LIVESTOCK AND POULTRY

Garden plants are not the only ones on the farm to suffer from temperature extremes. Livestock and poultry are also susceptible. Depending on where you live, measures must be taken to protect animals from the elements. In some climates, winter time is the danger zone, requiring warm, secure barns and heated watering devices that won't freeze over. In other areas, summertime is the most challenging season, in which animals are at risk of dehydration and heat exhaustion.

Poultry has particular seasonal considerations. Egg production can slow down or halt when hens experience cold or heat stress due to temperature extremes. Egg production also decreases when day length dips below fourteen hours of sunlight.

Here are some tips to keep livestock healthy and productive during seasonal extremes.

DEEP LITTER COMPOSTING

If you have any experience with composting, you may be familiar with the heat that is given off as microbes break down the materials in the presence of oxygen. This residual heat can be used to keep barns, coops and enclosures warmer, which is why many farmers opt for deep litter composting during the wintertime. Deep litter composting allows animal waste, hay and straw to build up to thick layers in the animals' living quarters. As the materials compost down, the pile gives off heat, keeping enclosures warm, reducing or eliminating the need for heating appliances. As a side benefit, wonderful compost is produced for the spring garden.

When deep litter composting is used, ventilation is important. Odors, dust and moisture must have a means of escape and fresh air should be able to freely enter the

enclosure in order to maintain a dry and healthy atmosphere for the animals.

Starting each fall, we begin to add straw and wood chip bedding to the floor of our chicken coop, mixing it into the old bedding and turning areas that have become matted. The chickens help to turn the materials as they scratch through it, and their waste ensures that the pile decomposes rapidly and stays hot. Goats, on the other hand, will not assist in turning the pile, and urine build up can give off unpleasant ammonia odors. To keep their enclosure smelling fresh, we use a garden fork to turn and fluff the materials when they become matted, and add dry materials regularly to the top of the pile. Learn more about deep litter composting by following the link in Figure QR 2.4.

THE DIRT
on farming
Deep Litter Composting
CITYFARMINGBOOK.COM/DEEP-LITTER-COMPOSTING/
QR 2.4

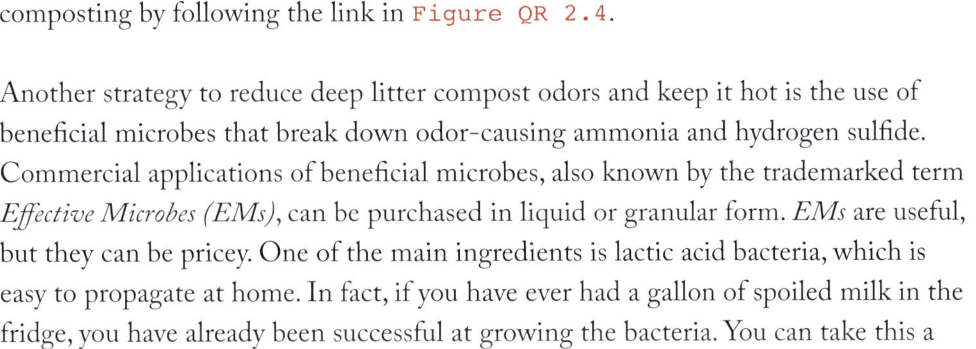

Another strategy to reduce deep litter compost odors and keep it hot is the use of beneficial microbes that break down odor-causing ammonia and hydrogen sulfide. Commercial applications of beneficial microbes, also known by the trademarked term *Effective Microbes (EMs)*, can be purchased in liquid or granular form. *EMs* are useful, but they can be pricey. One of the main ingredients is lactic acid bacteria, which is easy to propagate at home. In fact, if you have ever had a gallon of spoiled milk in the fridge, you have already been successful at growing the bacteria. You can take this a step further to create beneficial bacteria for composting.

HOME-GROWN BENEFICIAL BACTERIA

1. Pour 1 quart of milk in a glass jar and screw the lid on loosely. Store the milk in a dark cabinet until a curd forms on top of the liquid.
2. Remove the curd and add one tablespoon of blackstrap molasses to the liquid. The mixture can be stored in the refrigerator for up to one year.
3. To activate, mix one part liquid to twenty parts non-chlorinated, room temperature water. Spray or pour the mixture on to composting materials and turn in with a garden fork.

For more detailed instructions, following the link in Figure QR 2.5.

QR 2.5

RE-LEAF: PLANTING SHADE FOR LIVESTOCK

Just as the best type of shade for gardens and for humans is natural, the same is true for livestock. Deciduous trees and bushes provide shelter from the heat of the sun and have an evaporative cooling effect. As space permits, plant deciduous trees along an arc on the west, north and east sides of livestock enclosures. During summer heat, our chickens gather in the shade of the Desert Willow that we planted in their run, and they congregate under the bushes and fruit trees in the yard. Protected by the foliage, they stay cool and hidden from the hawks that fly overhead, looking for prey. In the safety of cover, they scratch and hunt all day for treasures of fallen fruit and mulberries. When the weather turns cool, bare branches continue to provide safety from hawks while allowing the hens access to as much direct sunlight as possible.

Sheep and goats benefit from natural shade, but they are prone to damage tree bark and limbs that are within reach. Goats in particular love to browse woody plants and prefer to eat shrubs and trees over grasses and weeds. Plant trees and large bushes outside of their enclosures and trim the lower branches to create a high canopy that throws shade into the pen. If you plant fast-growing varieties that are edible to goats and sheep, they will trim the branches that grow into the pen for you. They are happy to do this trimming work, and it can supplement their diet, reducing your feed bill. To find out more about edible shade and what types of plants are healthy for ruminants to eat (and what should be avoided!), follow the link in Figure QR 2.6.

SOLAR AND THERMAL TEMPERATURE REGULATION

When building livestock enclosures or coops, place the majority of windows and doors on the south side of the buildings for maximum heat absorption during the wintertime. Sunlight will shine into the building during the daytime, producing heat. Trap and store the heat by making sure that the building is well insulated and by keeping doors closed as much as possible. Also, ensure that trees, buildings and other obstructions do not block the sun from the south.

You can use a combination of solar heat absorption and thermal mass to keep livestock water buckets from freezing in the winter. One simple and inexpensive method is to insulate a water dish with a discarded tire. Stuff the interior of the tire with bubble wrap. Measure the diameter of the hole in the center of the tire, and procure a black, rubber water dish that will fit snugly into the hole. Livestock feed stores often carry multiple sizes of these dishes, and they can also be purchased online. Place a couple of bricks or blocks under the dish to create a layer of air between the bowl and the ground, thus reducing the amount of heat that is lost via conduction. Float a few pieces of wood in the water to break up any ice that starts to form. Situate the tire and dish in bright sunlight, where the black surfaces will absorb and store solar radiation to prevent the water from freezing.

We have found that keeping our water buckets in tires has a dual purpose, not only to keep the water warm, but also to prevent our sheep and goats from damaging the buckets. We have a delightful ram with a friendly personality. But being a ram, he likes to ram things, especially his water buckets. The tire serves as a bumper to keep the buckets upright and undamaged. View a photo of one of our tire bumpers and learn more about keeping livestock water from freezing by following the link in Figure QR 2.7.

HEAT LAMPS

Heat lamps are an easy way to raise temperatures inside coops and enclosures. For sheep and goats, we use a heat lamp only for kids and lambs. Mature animals can withstand cold temperatures, and I want them to develop their own defenses by growing a thick undercoat. But newborns do not regulate body temperature well and are susceptible to hypothermia.

Goats are extremely curious, and will damage any lamp that they can reach, so place them in a secure location. I hang them low, just outside the welded wire fence, angling the bulb to shine into the pen so that the kids can access the warmth without being able to touch the lamps. A red heat bulb seems to be more soothing to the animals at night than a bright, white bulb.

For poultry, heat lamps serve a dual purpose. They provide warmth and supplemental light that the hens need in order to boost egg production when daylight dips below fourteen hours. We use a white bulb that is plugged into a timer, which turns the light on at 4am, and switches it off in three hours. The logic behind this system is that chickens are able to withstand cold temperatures, but the most dangerous hours of the day occur just before sunrise. The heat lamp raises the temperature enough during the most frigid time of day to prevent stress and possible damage to the hens. The incandescent bulb provides just enough extra hours of light that they need to produce eggs and no more, so electricity is not wasted. The light comes on in the morning when the hens are naturally waking up, and does not hinder their natural instinct to return to the coop to roost at night.

Although we never experienced it, I have heard that extending the daylight hours in the evening will cause the hens to mill around longer on the coop floor. If the light goes out suddenly before they are roosting, they may have difficulty finding their roosts. A hen that cannot roost to sleep is an unhappy, stressed bird who may not lay, despite the supplemental light with which she is provided. In any case, having the light turn on automatically in the morning has worked well for us. And if there is any issue with the lamp, it occurs when we are heading out to do our morning chores rather than at night when we are less likely to notice any trouble.

Keeping a close eye on heat lamps is important, but they also must be carefully placed to reduce the risks associated with them. Heat lamps are easily broken by curious animals, and the bulbs become hot enough to cause a burn if they are touched. Additionally, if they come in contact with dry bedding, the bulbs and fixtures may be hot enough to cause a fire. Hang heat lamps well out of reach and away from flammable materials. Cage the fixture, if possible, to prohibit access and to prevent the lamp from accidentally falling. Follow the link in Figure QR 2.8 to learn more about managing the risks when using heat lamps.

HARDY LIVESTOCK BREEDS

Selecting breeds that are well adapted or native to your climate reduces the amount of management that is required to keep them warm or cool.

Chickens: Many of the chicken breeds that are prevalent in the United States today developed in the diverse climates of Europe, Asia and Africa and their traits vary widely according to their origins. Those that trace their roots to hot climates tend to be lightweight, thinly feathered and have large combs and wattles to wick

away heat. Cold climate breeds are often heavy with thick feathering and smaller combs. When selecting breeds for your farm, look for traits that will help them to survive and thrive in your particular climate.

As an example, many of our chickens are lightweight breeds, such as Leghorns, which tolerate the extreme Phoenix heat better than their heavier counterparts. Leghorns also have large combs and wattles that wick away heat. Another breed that we raise is the naked-neck, aka Turkens. These strange looking birds have no feathers on their necks, and the bare skin serves to disperse body heat and keep them cooler. Figure 2.8 shows one of our lovely Turken hens.

2.8

BREED SELECTION

CHOOSE TYPES THAT WILL THRIVE IN YOUR CLIMATE

ANIMALS THAT ORIGINATED IN A CLIMATE SIMILAR TO YOUR OWN WILL BE HEALTHIER, MORE COMFORTABLE AND EASIER TO MANAGE

On the other hand, if you live in a colder climate, heavy breeds, such as Jersey Giant and Brahma are extremely cold tolerant. Breeds such as Ameraucana, Buckeye and Chantecler have small combs and wattles that are less susceptible to frostbite and radiate less body heat. Some breeds have fancy feathering that helps to keep them warm. Select breeds with feathered feet, beards and muffs, such as Cochin and Faverolle. More information and resources concerning chicken breed selection are located at the link in Figure QR 2.9.

QR 2.9

Selecting Chickens for Your Climate

CITYFARMINGBOOK.COM/SELECTING-CHICKEN-BREEDS/

Goats and Sheep: Ruminants are highly tolerant of cold temperatures. Wool or hair grows thicker as the weather cools, offering them protection during the winter. Often, a three-sided structure with a roof to shelter them from wind and rain is adequate housing for adult animals. Our goats lived in igloo-shaped dog houses, pictured in Figure 2.9.

2.9

GOATS

TOLERANT OF COLD TEMPERATURES

A THREE-SIDED STRUCTURE OR DOG HOUSE CAN KEEP SMALLER BREEDS WARM AND DRY.

Though tolerant of cold weather, ruminants are easily stressed by heat. While any breed of goat or sheep can technically be raised in any geographic location, it is sensible to select breeds that are best adapted to the climate in which they will be raised. Location of origin is a clue as to their tolerance level for hot weather. For example, sheep breeds developed in the Caribbean and in Africa tend to have fine wool or hair that sheds in the summer rather than thick wool that requires shearing. At The Micro Farm Project, we raised Dorpers, bred for the arid regions of South Africa, and Barbados Blackbelly crosses. Neither breed required shearing, but our ram really appreciated a brushing to help remove his winter coat.

QR 2.10

THE DIRT on farming

Selecting Goat & Sheep Breeds

CITYFARMINGBOOK.COM/GOAT-AND-SHEEP-BREEDS/

In cold, wet areas, heavy meat breeds and those with long or thick wool may be better choices than hair breeds. To discover which breeds are best adapted to your environment, contact your local department of agriculture, university extension office, 4-H groups or Future Farmers of America chapters. More information and resources concerning ruminant breed selection are located at the link in Figure QR 2.10.

CONCLUSION

Garden and farm planning can be a very thrilling and enjoyable activity as you dream about what you will create. It can also be overwhelming as you consider all of the factors involved. In later chapters we will discuss other aspects, such as water access, seasonal considerations, and soil. But in order to keep things simple, I recommend creating a rudimentary map of your property, as suggested in the *Inventing Your Farm* sections. Start by penciling in potential locations for your garden and various farm elements with *only* the solar aspect in mind. You can erase and move them around later as your knowledge grows.

Urban Farming Keys:

1. Observe your property through the seasons to discover its microclimates. Take notes and get to know your property well before placing permanent gardens and other structures on your farm.
2. Fruiting plants require six to eight hours of sunlight each day. Full morning sun is ideal, but any sunlight will do.
3. In the Northern Hemisphere, gardens that face south receive the most sunlight throughout the year. Barns and coops with windows that face towards the south stay warmer in the wintertime, when unobstructed by shade.

4. Sunny and shaded microclimates can both be productive if you select the right plants for the right place.
5. Western exposures tend to be hottest, which is wonderful for plants that thrive in the heat. It can be detrimental to more tender plants.
6. Natural shade tends to be the most refreshing for garden plants as well as livestock. Trees, large bushes, vines and tall flowers are great options for sunny areas and the west side of a garden.
7. Plants and livestock breeds that are either native or adapted to your climate will generally be easier to manage than non-native varieties.

Thoughtful observation and placement of the elements of your farm (both plants and animals) can maximize its productivity while also reducing the amount of energy and work that are required. As you plan and grow your homestead, you will certainly make some mistakes, as I did at my first attempt to grow tomatoes. In many ways, learning to be an urban farmer is a matter of your enthusiasm holding up until your successes outpace your failures. Understanding the role of sunshine and temperature on your property puts you ahead of the game, and will go a long way to ensuring your farming success.

In the next chapter, we will explore water, how plants use it, how much they need, and various watering systems for gardens and livestock. We will also continue to develop your urban farm plan.

FEATURED FARM

MCC CENTER FOR URBAN AGRICULTURE

As interest in urban farming grows, colleges and universities are responding by providing urban farming and sustainable agriculture programs to meet the demand. One such program is Mesa Community College Center for Urban Agriculture. I visited with Peter Condon, Program Director for Urban Horticulture and Sustainable Agriculture. With him was Christee Rothbard, a land lab technician who was meeting the challenge of keeping plants and the fish in the aquaponics system alive through the intense heat of the Arizona summer. Although the temperature was 107°F during my visit, many plants were growing happily on the property. I picked up a few great tips from Peter and Christee for meeting seasonal challenges:

1. Use shade cover. During the summertime when temperatures are high and the angle of the sun is direct, grow tender plants under shade. Peter and I discussed tomatoes, and one of the mistakes that Peter pointed out that Arizona tomato growers make when planting their fall crop is "they get them in too early, but they don't protect them. And then, they shrivel up in the heat and the sun or they won't set pollen." I noticed a group of PVC pipes lying in a pile. Peter explained, "Those are little teepees that we put down over the roses and we cover our taller crops with them. With the shorter crops we put that really thin, white, floating row cover down. That helps to keep a lot of insects away, but also provides a bit of shade for our new transplants."

   Shade cover is useful for protecting plants from the sun's intensity, and is most helpful in reducing temperature. This is important not only in very hot climates for season extension, but also in milder climates with temperatures that typically stay below 100°F. Many cool season plants begin to bolt when temperatures rise into the low 80s. Plants in the Solanaceae family, such as tomatoes and peppers, will not set much fruit when temperatures rise above 90°F. Learning the temperatures at which your favorite crops thrive, and providing shade to mimic those temperatures, can help many plants to grow for a longer period of time and be more productive. Figure 2.10 shows a shaded area of the farm at MCC.

GROW IN ANY CLIMATE

LEARN WHICH VARIETIES THRIVE IN YOUR CLIMATE. PLANT THEM IN THE RIGHT PLACE IN THE RIGHT SEASON.

2.10

2. Harden off transplants before planting. MCC has a greenhouse in which plants are started. Before transporting mature starts from the greenhouse to the community garden, they are hardened off under partial shade cover to help them to adjust to the outside world. For home gardeners, plants that are started indoors or purchased at a nursery should also be hardened off before they are placed in an outdoor garden. Giving transplants a partially protected environment helps them to adjust to wind by strengthening stems and branches. It also helps them to acclimate to increased temperatures and evaporation by thickening the waxy coating on their leaves. Spending a few days or a week in a semi-protected area can help to reduce transplant shock and increases the likelihood that transplants will survive.
3. Get a greenhouse. Peter recommends getting a greenhouse, if you are able, that will provide temperature control. "That really helps with getting a jump on winter starts and early spring starts." In climates characterized by temperature extremes, a greenhouse allows a gardener

to start plants much sooner than they can be started outdoors. Speaking of growing transplants for the fall, Peter says, "If you want to get tomatoes ready for harvest before the freeze, you really want to have (the transplants) a pretty good size and already in the ground ..." so that once the sun intensity dies down, "... you've got a plant that's ready to produce. So you'll have gained already a month or two." In a cold climate, spring plants can be started during the wintertime for transplant during the growing season. If you cannot afford or do not have space for a large greenhouse, backyard-sized greenhouses, cold frames, or even a sunny window will also give the gardener a jump on the seasons.

4. Work the soil during the off-season. In Arizona, many gardeners do not grow plants during the summertime, due to the intense heat. "The trick is," Peter says, "when it's really hot like this, you get out (in the garden) before work. This is when you work in your soil, when it's too hot to plant. Then when it comes planting time you don't start from scratch getting all the weeds out." In cold climates, the off-season may be in the wintertime. Composting, planting cover crops or root crops, and covering the soil with straw are off-season activities that can help to condition and prepare the soil for the growing season.
5. Use summer heat for solarizing. As we toured the MCC community garden, Peter pointed out, "Over here we've got another gardener who's solarizing their plot. They've had this black plastic on for probably two months now and they're trying to let Mother Nature, through her solar generator, kill the Bermuda grass and theoretically kill any weed seeds that are in the weed bank in the soil. It takes a few months." Solarizing is a process of using the sun to sterilize soil. It is accomplished by covering the ground with black or clear plastic, which is held firmly to the earth at the edges in order to trap heat. This creates a greenhouse atmosphere under the plastic where temperatures rise to levels in which weeds and grasses don't survive. Solarizing is useful for clearing ground of invasive grasses and weeds. However, the heat also kills soil microbes. Use with caution only as a drastic measure. And be sure to build the soil back up with plenty of composted organic matter prior to planting.
6. Grow plants that thrive in your climate. In the MCC gardens, basil and rosemary were growing very well in the heat. "Basil loves the summer heat," Peter comments. Eggplants and peppers also dotted the landscape. Seasonality is important. "We do teach students to follow planting

calendars, follow proper crop rotations, so they can grow all year long." Planting cool and warm season vegetables at the proper time is key to production, whether your climate tends to be hot, mild or cool.

THE DIRT
on farming

MCC
Full Interview

CITYFARMINGBOOK.COM/MCC/

QR 2.11

Peter encourages growers who live in harsh climates to give gardening a try, despite the challenges. "If you follow the best practices, you could farm all year long; you can make a living. It just takes a lot more practice; it takes more research, and just getting out there."

Read the full interview with Peter and Christee by following the link in Figure QR 2.11.

Chapter 3

Soak It Up!

Watering to Promote Vigorous Plants, Healthy Animals and a Flourishing Environment

My personal joy at seeing gray skies once again was interrupted by the guilty thought that, perhaps, someone should call random rooms at local resorts to apologize. It was the fourth day of rain in the valley, a boon to residents. But for all of the visitors expecting to escape the dreary winter weather in their home states and enjoy some Arizona sunshine, I felt badly … though not too badly.

It was the kind of weather that Phoenicians savor. Cloudy days are so few and far between that we are known to post on social media anticipatory videos of monsoon storms rolling into the valley and delighted rain-drenched selfies. Our friends in Portland or Seattle roll their eyes and book their flights to Phoenix in February to soak up some of our excess sunshine.

Today, the unusual weather was altering my watering plan for the better. In an unprecedented moment of opportunity, I pushed the rain delay button on my automatic irrigation system for the fourth day in a row, refreshed by the smell of the rain and the rare blessing of free water.

Prior to moving to Arizona, I had lived all of my life in the humid climates of the Midwest, Thailand, East Texas and the South Pacific. As the daughter of a water resources engineer, passionate about his projects, my eyes have been trained since childhood to take note of dams, culverts and retention ponds. It seemed to me that there was water everywhere; too much water that had to constantly be harnessed, redirected, contained, or removed. Groundwater was plentiful, and our well in Iowa was sweet. With all this available water, conservation and water shortages never crossed my mind.

Then, in 1993, I moved from the humid South Pacific to Arizona. As the plane descended through a blanket of brown, dusty air, the city of Phoenix came into view, also brown and dusty. Green desert trees and blue swimming pools punctuated the landscape. The appearance was like the negative image of typical city, the opposite of the predominantly green and blue ocean landscapes, dotted by occasional brown patches, to which I was accustomed. Exiting the plane, hot, dry air drew the moisture from my eyes, my sinuses, my lips. This was my first taste of desert living, but I would not discover the depths of what that really meant until years later when I became an urban farmer.

Once again, tomatoes were my motivation and my guidepost. That first season when the tomato harvest had finally materialized, something not so pleasant also

materialized – a $200 water bill. In Phoenix, every drop of water is precious, much of it shipped to us from wetter states via an elaborate network of rivers and canals. What I didn't realize was how expensive that water could be and I felt the pain of it when I wrote the check.

I was fairly certain that we had not grown $200 worth of tomatoes to offset the cost of the water. Though I enjoyed the fresh tomatoes, I wasn't willing to go broke to grow them. I began to ask the question, *Exactly how much and how often should I water my tomatoes?* I was looking for the minimum amount of water needed to grow the maximum number of tomatoes. The answer, I discovered, is *it depends*.

It depends is a response that I find myself using regularly, and it originally came from my good friend Greg Peterson of UrbanFarm.org. At the time, although delivered with much sincerity and kindness, it sounded flippant. But as I dove into the topic more deeply, I discovered that while my query seemed simple, it was a loaded question, one that demanded more questions in order to formulate an accurate response. Temperatures, soil types, plant varieties, humidity and, of course, rainfall, are a few of the factors involved.

Whether you live in a dry climate or one that receives lots of rainfall, this chapter will help you to discover how plants use water and the factors that will affect how much and how often you will need to water your garden. Once you understand these factors, you will be able to perform what I call *precision watering*, applying the right amount of water at the right time to optimize the health and productivity of your plants without wasting this precious resource.

We will also touch on water harvesting techniques that make optimal use of water, whether you have limited irrigation resources, as we have in Phoenix, or whether you have moisture in excess. Finally, we will discuss simple watering systems and how to make use of harvested water for livestock.

WATER WISE

In order to determine the most effective and economical watering plan to your garden, it is important to understand a few guiding principles.

WATER THE SOIL

The first principle is to water the soil, not the plant. Although overhead sprinklers are common for watering lawns, they are not ideal for garden plants. Regularly wetting leaves can contribute to fungus and mildew problems, not to mention the large water loss due to evaporation. Instead of spraying your garden from overhead, direct the water precisely where the plants will use it, at the soil level.

Besides sprinklers, which we have ruled out, the two most common ways to water a garden are by hand or via a drip irrigation system. The advantages to watering by hand are that it provides a regular time when you will visit your garden. This gives you an opportunity to discover more readily when issues pop up and when plants are ready for harvest. Many gardeners enjoy hand watering and find it relaxing. If you choose to water by hand, direct the spray to the soil level and avoid showering the leaves.

An alternative to hand watering is drip irrigation. Drip irrigation is a more precise watering method than hand watering, delivering water directly to plant roots. It is especially useful for large gardens, for which hand watering would be very time consuming. Figure 3.1 shows drip irrigation lines in a raised bed at The Simple Farm in Phoenix, which is a sizable urban farm with too many beds to water by hand.

3.1

DRIP IRRIGATION

DELIVERS WATER AT THE SOIL LEVEL

DRIP IRRIGATION CONSERVES WATER AND HELPS TO PREVENT WATER-BOURNE PLANT DISEASES, SUCH AS MILDEW AND FUNGAL INFECTIONS

In an arid climate like ours in Phoenix, hand watering can become a time-consuming chore. Because we have to water so often during the dry months, I prefer to use a drip irrigation system. Drip irrigation applies water slowly, directly into the soil, via

a network of water emitters. These systems are highly efficient, and when combined with a timer, they run automatically. This can be a life saver in climates where one day of missed or forgotten watering can spell disaster for the garden. If you choose to use an automatic drip system to water your garden, check equipment regularly to make sure that it is running properly and to spot any maintenance issues that arise. Discover more about drip irrigation and maintenance at the link in Figure QR 3.1.

HOW WATER MOVES THROUGH SOIL

The second principle is to understand percolation. You may be surprised to learn that once water infiltrates soil, not only does it travel through the soil vertically, drawn downwards by gravity, but also horizontally, much like the way the water moves through a sponge. How quickly the water moves and how widely it spreads out are largely determined by soil texture. Soils that are high in clay content initially resist water infiltration. But once water penetrates the surface, clay soils tend to hold onto moisture, dispersing it laterally with limited downward flow. Due to their tendency to absorb moisture and to drain moisture slowly, it is generally recommended to water clay soils infrequently with a long, slow drip.

In contrast, highly porous sandy soils absorb and disperse water quickly in a predominantly downward direction. Due to their fast draining quality, sandy soils are irrigated frequently with quick, small bursts of water.

To discover whether your soil has a high sand or clay content, dig out a handful and get it wet enough for the particles to hold together. Roll the soil back and forth between your hands to create a ribbon. If you are able to make a long ribbon out of the materials, the clay content is probably high. If the ribbon falls apart easily, the soil most likely has a high sand content. For more details about soil structure and texture, follow the link in Figure QR 3.2.

WATER AT THE RIGHT TIME OF DAY

The third principle is to water at the right time of the day. Conflicting information circulates concerning this topic. Some gardening experts advise watering at dusk or during the night, while others recommend dawn or daytime watering. A brief botany lesson will clear up the confusion. Plants uptake water via a process called *evapotranspiration*, by which water in the soil is pulled into the plant from the roots, upwards through plant tissues, and finally evaporated from pores in the leaves called stomata. This movement of water through the plant transports minerals, keeps the plant upright, and has a cooling effect. Evapotranspiration uses approximately 90% of the water that enters the plant. The other 10% is reserved for photosynthesis and cell growth.

With the exception of succulents (including cacti) most plants perform evapotranspiration and photosynthesis during daylight hours, closing their stomata and shutting down the processes at night. Since typical garden plants use water during the day, it makes sense to soak them when they are active. Even so, some will argue for night time irrigation by pointing out that watering when the sun and temperature are up increases the amount of water loss due to evaporation. My recommendation is to water just prior to sunrise, typically the coolest and most humid time of day, early enough for the moisture to penetrate the soil prior to sunup. By doing so, you will diminish water loss due to evaporation, minimize fungus and mildew growth that thrives in standing water, and maximize the number of hours that your plants are able to make use of the water.

WATER AS DEEP AS THE PLANT IS TALL

The fourth principle is to water as deeply as the plant is tall. Plant roots are roughly a mirror image of the foliage that grows on top, and the root tips are the most active part. In order to grow a strong, healthy plant, it is important to provide moisture to the entire width and depth of the rootball. Low growing lettuces and herbs may require only a 10in depth of moisture in the soil, while a taller tomato bush may need 2ft of water penetration.

Whether you water by hand or use a drip irrigation system, you can calculate how long you will need to water in order to reach the appropriate depth using a soil probe, a moisture meter or a long screwdriver. At first, you may not have any idea how long it will take for moisture to reach the desired soil depth, so you will have to make a guess. You might start by running the water for fifteen minutes, after which you will turn the water off and allow it to soak into the ground for a half hour. Then, push the probe gently into the soil. It will slide easily where the water has penetrated. Once you reach dry soil, you will feel a change in resistance. If moisture has not reached the appropriate depth, run the irrigation for an additional fifteen minutes, again waiting thirty minutes for percolation and retesting with the soil probe. Repeat the process until you can easily slide the probe to the desired depth, keeping track of the time that it takes to achieve. This is how long you will need to hand water or set the timer for your irrigation system to run.

When calculating irrigation for the first time, you may be surprised, even alarmed, by the amount of water required to penetrate the soil deeply, especially if your garden has a high clay content. For gardeners in arid climates, it may seem wasteful to run the irrigation for such long periods of time. Although it may be counterintuitive, deep irrigation is a method of water conservation that sequesters moisture down deep in the soil, reducing evaporation and increasing the number of days between irrigation sessions. Plants will be healthier, with strong root systems, and will need to be watered less often. Additionally, in areas that have high mineral content in the water, salts can build up in the soil, causing drought-like conditions for plants by holding on to moisture. Deep watering flushes damaging salts away from the root zone so that plant roots can readily uptake available soil moisture. Access a watering guide at the link in Figure QR 3.3.

WATER ONLY WHEN NECESSARY

The fifth principle is to water only when it is necessary. This may seem obvious, but the truth is that over-watering and under-watering have similar damaging effects, and it is easy to succumb to the temptation to irrigate too much. Once you know how long to run your irrigation system to reach the recommended depth, you will need to determine how often to run the system in order to avoid over- or under-watering. There is no standard answer to this question. A good rule of thumb is to water deeply, and then refrain from watering until the top 2 or 3in of soil are nearly dry. The best way to determine when it is time to water is to stick your fingers into the soil to feel for moisture content. I personally do not rely on moisture meters, which can be unreliable. Fingers are free, readily available, and accurate. If soil feels cool and moist, refrain from watering. If it feels dry or nearly dry, water.

You may be wondering how to use an automatic irrigation timer if the gardener has to feel the soil in order to know when to water. The answer is that when you are first starting out, you will need to monitor the soil moisture levels and keep track of your watering sessions. After several seasons of gardening, it will become more intuitive. As an example, I know from years of gardening that I will need to water every other day during May and June, every third day during the monsoon season, and once or twice per week during the cool months. It is not a precise system, but it allows me to reset my timers to water the appropriate number of days each week for the season. Of course, if we receive more rain than usual or experience a particularly hot, dry spell, I can delay the timer or run an additional irrigation session as needed.

INSPECT BEFORE YOU WET

The sixth principle is to inspect before you irrigate. I mentioned in the previous section that over- and under-watering have similar damaging effects, and they can have similar symptoms. For example, a plant may wilt because the soil is dry, or it may wilt because the soil has been too wet for too long and the roots are essentially drowning. Repeatedly watering a wilted plant without checking the soil conditions may eventually result in the plant's demise. This is a common occurrence with house plants which, in my experience, are more often killed by too much attention and watering than by benign neglect.

Another example worth noting is heat induced wilting that can occur on hot afternoons, even in well-irrigated gardens. When temperatures rise, evaporation can outpace the ability of roots to uptake water, resulting in a drop in water pressure that causes stems to go temporarily limp. The temptation is to assume that the plants need water ASAP. A wise gardener will feel the soil before turning on the irrigation. If the soil is dry, water may be necessary. But if the soil is moist, irrigating only results in wasted water that will not remedy the situation. If soil moisture seems to be adequate, refrain from watering. You will likely notice that the plants will perk up when dusk falls and the temperature wanes.

TURNING WATER CHALLENGES INTO ASSETS

ORGANIC MATTER

When gardeners discover that they have an extreme soil type, whether it is sandy or high in clay, they will often ask how to improve their gardens so that they either have better moisture holding capacity or better drainage. The remedy in either case can be to condition the soil with organic matter, generally in the form of compost. Whether you purchase bagged compost or make it yourself, several inches of composted material tilled into the top 6in of sandy soil will increase its water holding capacity as much as 250%, reducing the frequency with which you will need to water. When added to clay soils, compost creates spaces that allow for better drainage and air penetration. More information about improving both droughty and boggy soils with organic matter is at the link in Figure QR 3.4.

MULCH

Mulch is a layer of organic matter that is spread on top of garden soil like a blanket. Compost, wood chips, bark, pine needles, dry leaves, hay and straw are common mulch materials. Mulch serves a number of wonderful purposes, from suppressing weeds and regulating soil temperature to attracting earthworms and providing nutrients. It is also extremely useful at reducing water evaporation from soil, thereby improving the garden's water holding capacity. Additionally, mulch aids the infiltration of surface water through the top layer of soil so that when it rains or when you irrigate, the water will penetrate the soil more readily instead of pooling or running off. Figure 3.2 shows wood chip mulch being spread in a garden.

ORGANIC MATTER

COMPOST, MULCH, & MANURE

INCREASES WATER-HOLDING CAPACITY OF SOIL

REDUCES EVAPORATION AND RUNOFF

3.2

One important point concerning mulch is that, with the exception of compost, mulch is used as a blanket on top of the ground and should never be tilled into the soil. The reason for this is that plants require nitrogen to grow. As mulch materials are decomposing, the microorganisms that perform the process also require nitrogen, which they will rob from the soil for their own use during the breakdown process. Wherever the mulch touches the soil, there will be almost no nitrogen available to plants. Therefore, it is best to keep mulch at the surface of the ground, and not down deep where plant roots are feeding. When mulching around seedlings or small transplants that have shallow root systems, application of organic nitrogen fertilizer prior to laying the mulch will supplement nutrient levels until roots have grown long enough to find the nitrogen that they need at deeper soil layers.

THE DIRT on farming

Mulch

CITYFARMINGBOOK.COM/MULCH

QR 3.5

View photos and read more about types of mulch that we use at The Micro Farm Project by following the link in Figure QR 3.5.

NATIVE PLANTS

When selecting plants for your garden, explore varieties that are native or adapted to your area. These plants have many benefits, one of which is the ability to survive only on natural rainfall, requiring little or no irrigation. In arid climates, low water use plant varieties are abundant, from succulents and cacti to drought-tolerant trees, grasses and flowers. Though typically grown for their beauty, many of these ornamental plants have an edible or medicinal element. At The Micro Farm Project, we grow cacti that have edible fruits and pads, as well as native Palo Verde and Mesquite trees that produce edible beans. Figure 3.3 shows fruits harvested from a Cereus cactus that is in our front yard.

3.3

Desert plants require no irrigation, once established. In our vegetable garden, we include drought-tolerant varieties of beans, squash, peppers, flowers, berries and melons. A helpful resource that we have found for seed and education concerning varieties adapted to arid climates is Native Seeds, which can be found online at www.nativeseeds.org.

QR 3.6

THE DIRT
on farming

Growing Native Plant Varieties
CITYFARMINGBOOK.COM/NATIVE-PLANTS/

In wet climates, native plants will thrive in boggier conditions, reducing runoff and soil erosion. They tolerate fluctuations in moisture levels, withstanding dry seasons and periodic drought. No matter where you live, use the internet or contact your local Master Gardeners for a list of native and adapted plants for your area. The link in Figure QR 3.6 will point you to a few online resources.

SLOPING TERRAIN

Many properties have fluctuations in terrain. Though we don't often think of slopes as ideal areas in which to grow a vegetable garden, they can be very productive with proper plant placement. Water runs naturally from high to low areas via gravity, collecting at the lowest point. Therefore, when gardening on a slope, drought-tolerant plants can be placed at the highest point, reserving the bottom of the slope for plants that thrive in wetter conditions. Growing plants on a hillside or slope is not only beautiful, it also has the effect of reducing erosion as plants slow the flow of water and roots hold on to the soil. View some photos and tips at the link in Figure QR 3.7.

THE DIRT on farming

Slopes & Hillside Gardens

CITYFARMINGBOOK.COM/HILLSIDE-GARDENS/

QR 3.7

MAN-MADE LANDFORMS

Swales and basins are man-made depressions in the earth that slow the flow of storm water runoff so that it infiltrates the soil. In arid regions, they are used as means of passive rainwater collection. Swales and basins are often planted with drought-tolerant varieties that can survive on storm water alone, or with very little supplemental irrigation. Swales are sometimes combined with curb cuts that allow water to run off of a roadway into depressions that are dug and planted along a sidewalk or in a median. The collected water is used to grow plants that beautify the roadway, as well as for shade or fruit trees.

Swales connected to curb cuts are not recommended for vegetable gardening, due to the high number of toxins that may be present in water that runs off of the street. However, if your property is sloped, you may use swales and basins in a similar manner to catch and slow rainwater that runs downhill or off of your roof for use in your garden. If your property is flat, digging your garden bed below grade will form a basin that collects rainwater and sinks it into your garden bed. It is also possible to direct roof runoff, greywater or air conditioning condensate to the basin to provide additional moisture. Basins have the added benefit of holding deep mulch, which keeps the water in the soil by slowing evaporation. See tips for desert gardening and sunken gardens at the link in Figure QR 3.8.

QR 3.8

WATER HARVESTING STRATEGIES FOR WETLANDS

QR 3.9

SWALES AND SUMP PITS

Although swales are primarily a passive water harvesting tool that is implemented in dry climates, they can also be used to reduce flooding and erosion in wet climates. Swales passively direct the flow of storm water using landforms and gravity, reducing how quickly it spreads and sinking it back into the soil. They can also direct water into a dry well (aka sump pit) or a garden bed. Additionally, they are helpful on sloped properties to direct runoff water, preventing the formation of gullies and nutrient loss due to erosion. There are so many applications for swales and sump pits, and resources are located at the link in Figure QR 3.9.

ORGANIC MATTER

The combination of a wet climate and compacted or clay soil can result in a waterlogged garden. Plant roots must have oxygen to survive, but chronically wet soils resist necessary air penetration. Prolonged waterlogged conditions generally lead to languishing plants. During wet seasons, good drainage is vital to garden health and vitality.

To test drainage, dig a hole in your garden to a 1ft depth and fill it with water. If the water fails to drain within an hour, the soil has a drainage problem that may affect plant growth. One of the best and easiest ways to improve drainage is to amend the soil with organic matter. Use a garden fork to work 1in of compost into the top 6in of soil, and cover the soil with 1 or 2in of organic mulch. Repeat this process at the beginning of the growing season when you are planting, and again at the end of the season to prepare it for dormancy. When drainage is particularly poor, you can initially turn several inches of compost into the soil using a tiller. Refrain from repeated tilling, however, as it may damage long-term soil health.

If you garden continuously with no dormant period, as we do in Phoenix, make applications of compost part of your planting routine, working compost in with a garden fork or adding it to planting holes.

RAISED BED GARDENS

If drainage or flooding are problematic in your landscape, a raised bed garden may be the ideal solution for growing vegetables in excessively moist conditions. Figure 3.4 shows raised beds that are elevated above the ground. If you cannot elevate a bed, place the bed on top of the wet ground, and fill with well-draining soil, such as potting soil that contains sand, perlite or pumice to improve drainage. Raised beds are typically made of lumber. If you are building the bed in a wet area, use the hardest wood available, such as redwood, to prevent rotting. Use 2×4 or 4×4 board dimensions. Thinner boards will quickly warp and crack in moist conditions.

3.4

RAISED BEDS

GARDEN SOLUTION FOR EXCESSIVELY WET CONDITIONS

BEDS CAN BE PLACED ON TOP OF WET SOIL, OR ELEVATED ABOVE THE GROUND TO INCREASE DRAINAGE

Protect the wood with an organic or natural wood sealer. Many options are available on the market, or you can make your own. My recommendations for safe wood sealers are located at the link in Figure QR 3.10. Alternatively, raised beds can be assembled from rock, cinder blocks, bricks, metal, cement and many other types of materials.

QR 3.10

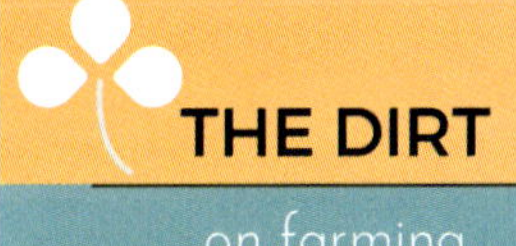

on farming

Safe Wood Sealers for Raised Bed Gardens

CITYFARMINGBOOK.COM/SAFE-WOOD-SEALERS/

THE DIRT

on farming

Sheet Mulching/
Lasagna Gardens

CITYFARMINGBOOK.COM/
LASAGNA-GARDENING/

QR 3.11

SHEET MULCHING

One of the easiest ways to build a raised bed garden without digging is sheet mulching. Also known as *lasagna gardening* or *composting in place*, the idea is to build a raised bed garden frame on top of the existing soil and fill it with compostable materials. As the organic matter breaks down, composted soil is created within the frame. Meanwhile, the activities of microorganisms and other decomposers in the compost serve to condition the native earth underneath the bed. Moisture-holding capacity, drainage, and soil fertility are improved without the intensive labor associated with digging and turning soil. View photos of how we assembled sheet mulch beds at our farm and resources for creating your own lasagna garden at the link in Figure QR 3.11.

RAIN GARDENS

A rain garden or pond garden can be a beautiful and useful addition to your landscape to soak up excess water. Similar to a swale, rain gardens are shallow depressions in which water can collect during a storm and slowly percolate into the soil. Placed near a runoff source, rain gardens are an inexpensive way to reduce flooding and put excess water to use.

3.5

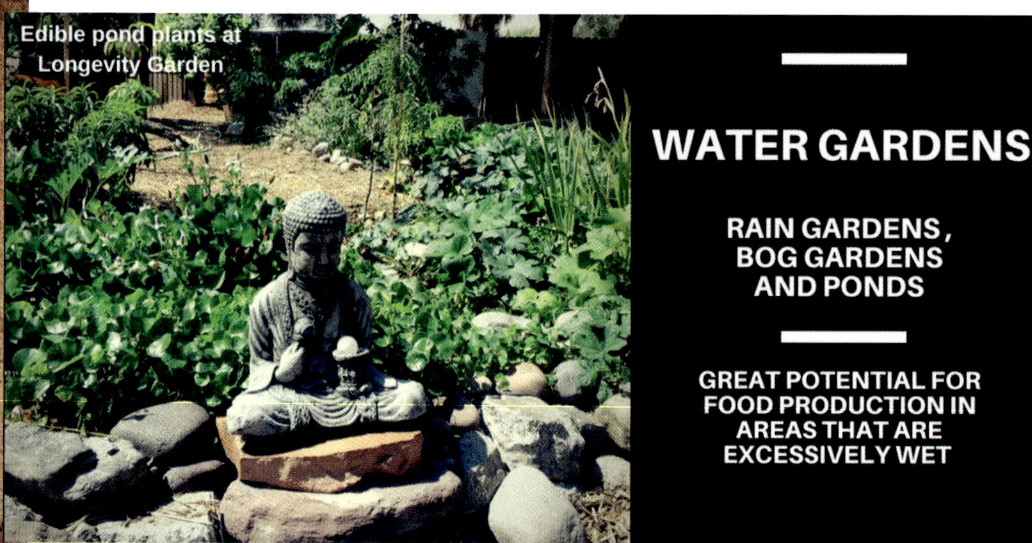

Edible pond plants at Longevity Garden

The best plant varieties for a rain garden are native plants that require minimal care and watering between downpours. The edges of a rain garden will dry out more quickly than the middle, so plants with the highest tolerance for wet soil are planted in the center where the earth remains submerged for the longest period of time. Most rain gardens are planted with grasses and flowers, selected for their beauty and hardiness, as well as to attract pollinators. Figure 3.5 shows a lovely pond garden planted with water-loving plants.

QR 3.12

Rain gardens can also be used to grow edible plants, with some caution concerning pollutants that may flow into the garden with the storm water. Because toxins may accumulate in the leaves and roots of plants, select edible plants that fruit rather than leafy greens or root vegetables. More information about rain gardens, bog gardens, and edible water gardens are located at the link in Figure QR 3.12.

STRATEGIES FOR POULTRY AND LIVESTOCK

SIMPLE SYSTEMS TO PROVIDE SAFE, CLEAN WATER WITHOUT A LOT OF EFFORT

Fresh, clean water is a necessity for livestock. It is basic to supporting life, and when livestock experiences even mild water stress, it can inhibit milk and egg production. The daily task of providing clean water is critical, and it can be challenging. During the wintertime, water sources are in danger of freezing, creating a cap of ice that prevents animals from drinking. And when temperatures are warm, algae grows and water quickly becomes putrid. Wind, direct sunlight, freezing temperatures, and the animals themselves can damage watering devices, requiring a vigilant eye and ongoing maintenance on the part of the farmer. Though it may not sound like a daunting task, daily water chores can become tedious, even overwhelming, especially on farms that raise several species of animals.

My philosophy is that an urban farm should improve my health and well-being, and it should be as simple and fun as possible. To those ends, we try to minimize and streamline tasks that induce daily stress, which includes watering. In Phoenix, water is

of utmost importance, particularly in the summertime when a failed watering system can result in animal death within hours. We have tried many different systems, and have discovered a few guiding principles that make watering as easy as possible. There is no one way that is best to provide water in every climate and situation. When considering a watering system for your farm, follow these guidelines to keep the process simple:

1. **Automatic:** Systems that provide water automatically can take the daily pressure off the farmer to refill watering devices. When we first started out with chickens, we used jug watering systems that had to be cleaned and refilled daily. While these systems were inexpensive to purchase, they were awkward and difficult to refill, and too heavy for my children to carry. We eventually had a water spigot installed next to the coop in order to attach an automatic watering device. The ease afforded by retooling the system was worth the upfront expense, and I highly recommend installing an automatic device for livestock.

   There are two main types of automatic watering devices. Systems with floats sense when water is low and automatically refill, turning the water off when the container is full. Another option is a system that animals can learn to operate, releasing water by pressing a lever. Some examples are chicken nipples or cups, which may sound funny, but are very handy in the coop. Chickens quickly learn to operate these devices to release water only when requested.
2. **Closed:** Water troughs, buckets and dishes that are open to the environment quickly collect debris. They are contaminated by the animals themselves, which may step in or perch on the containers. Chickens are particularly notorious for standing in open water sources, defecating in them and spoiling their drinking water. A hood or cover placed over a water source is a minimal prevention measure. Fully enclosed systems are best to prevent environmental contamination, and they also serve a side benefit of reducing water loss due to evaporation.
3. **Electricity-Free:** Because electricity is costly and can fail, I prefer to use systems that do not require power. In climates that are cold enough in the wintertime to freeze watering devices, electric heaters may be necessary. But if the heater fails, animals will not have access to water and the equipment may break. When possible, I recommend using power-free methods to keep water from freezing. For example, black buckets placed in the sun may be enough to provide ice-free water during the daytime.
4. **Easy to Clean and Maintain:** To minimize the time needed to perform routine maintenance, select a system that is easy to take apart for cleaning. Because animals can be very hard on the equipment, I also prefer devices that are

very sturdy and have as few working parts as possible to reduce the possibility of system breakdown. Additionally, I look for systems that have inexpensive and easily replaceable parts, and I keep a small stock of these parts on hand for easy and quick repairs. Figure 3.6 shows our daughter caring for a friend's miniature horse, for which watering was easy due to a simple spigot and hose system that was located adjacent to the enclosure.

3.6

5. Redundant: Provide a backup to your system. This is especially important in hot, dry environments, but also applies in milder climates. Should a system fail or a water source become contaminated, having a backup can save your animals' lives.

WATER HARVESTING FOR LIVESTOCK

Coops and pens are often located far from a water source. This was the case at The Micro Farm Project before we had water spigots installed in more convenient locations. Hauling buckets and hoses was quite a chore, and the water was costly. In order to alleviate the pressure on our time, energy and budget, we installed a water harvesting system to catch air conditioner condensate for our chicken coop. On a daily basis in the summertime, I estimate that 5 to 15 gallons of water were dripping off the roof in an area within 30ft of the chicken coop. Instead of allowing the free, clean water to evaporate, we began to collect it as a backup water source for our birds. The proximity of the collection tank to the coop made an easy task of rinsing and filling water dishes, and I could connect the tank spigot to a water dish with a float to automate the system.

If we lived in a climate that received more rainfall, I would also install gutters to our chicken coop and animal enclosures to facilitate rainwater harvesting. In a storm, even a small roof can shed enough water to fill a rain barrel or two. This is enough water to supply livestock for several days, located conveniently near water troughs and dishes. The tanks could also be connected directly to an automatic watering system. Follow these guidelines to ensure that the water is safe to drink.

Keep It Clean: Roof contaminants, such as bird droppings and leaf litter, can make collected water unfit for animals to drink. A *first flush* diverter will remedy the situation. When rain begins to fall, this device automatically redirects the first wash of dirty water that enters a downspout, diverting it away from the collection container. Following the first flush, cleaner water is allowed to enter the cistern.

Keep It Dark: Locate collection tanks out of direct sunlight to prevent algae from growing in the stagnant water. Algae grows only in the presence of light. If your tank is translucent, you will want to cover it or paint it black to keep the water and the spigot clear of algal growth.

Keep It Covered: Closed water systems prevent mosquitos from using your stored water as a breeding ground. Seal any openings with spray foam or screening.

Keep It Legal: Check with your local government to make sure that rainwater collection is allowed in your municipality. In some states, water rights belong to the government, and rainwater harvesting is not allowed or is restricted to specific uses.

Keep It Passive: In order to make it as easy and economical as possible to water your livestock, design your water harvesting system so that it delivers the water using gravity. Elevate collection tanks or dig trenches that decline from the water source down to the point in which it is dispersed to the animals. A 1% grade, or ⅛in per foot, is enough of a decline to transport water without the use of any pumps or mechanization.

More information about water harvesting, first flush diverters, and water harvesting legalities is located at the link in Figure QR 3.13.

CONCLUSION

Cartoonist Lou Erickson famously quoted, "Gardening requires lots of water – most of it in the form of perspiration." Keeping a garden moist enough, but not too soggy, can be challenging in several ways, from knowing how much water is enough to the investments of money and time that it requires. Fortunately, once we understand how plants use water, we can design irrigation systems that meet their needs. And nature gives us a hand. I have heard it said that *God made rainy days so that gardeners could get the housework done*. Though humorous, the saying illustrates that tap water is not the only source of water available to gardeners. Even in arid climates, water can be harvested from the landscape for use in the garden. Water harvesting systems not only conserve water, but if they are designed using gravity to passively direct the flow, they also conserve labor on the part of the gardener.

We have learned how plants use water and best practices for watering a garden. We have also discovered how to conserve water and a few techniques to overcome water challenges. Finally, we touched on using rainwater to offset the cost and labor associated with watering livestock animals. By using water properly, we can create the healthiest, most productive urban farms while conserving one of our most precious natural resources.

Next, we will explore growing seasons and season extension, as well as how seasonal changes affect livestock production.

FEATURED FARM
CRAIG'S GARDEN

My good friend Heather of Taylor Turned Earth lives in a nifty neighborhood in the Arcadia district of Phoenix. Many of the homes are being renovated to look like farm houses. The yards have flood irrigation, a watering method that is unique to the area and that is a remnant of our agricultural days. Gardens flourish in the neighborhood, many of them located in front yards. Heather and I are working on a project together, and she wanted me to see one of her neighbor's gardens. We took a photographer friend along with us to take pictures of the property, which utilizes a number of water-saving strategies.

During the visit, I had the opportunity to briefly meet the gardener, Craig, who pointed out a few highlights from his garden. First, and perhaps most importantly, the garden is sunken below grade to retain and infiltrate moisture into the soil. Figure 3.7 shows an example of a sunken bed in Craig's garden. Berms create a waffle pattern in the ground so that he can direct more water to plants that are thirsty, and less water to drought-tolerant plants. The low grade of the gardens provides a slightly cooler environment for the plants, reducing both heat stress and evaporation. Additionally, Craig creates dense plantings. Plants growing close together cover the soil, shading it and further decreasing water evaporation. Finally, Craig cultivates tall plants on the west side of his garden to provide afternoon shade to lower-growing vegetables.

Although flood irrigation is available, Craig does not use it due to the possibility that canal water may bring contaminants into his garden. He prefers to have more control of what enters his property and comes in contact with his vegetables. Neither does he use a drip irrigation system. Instead, he simply fills the sunken beds periodically using a garden hose. Craig's garden design and management practices significantly reduce the number of times that he has to provide supplemental water to his garden, an extremely important feature in the drylands of Arizona.

3.7

In wetland environments, sunken beds can be used to eliminate the need for supplemental irrigation. They can also control flooding and pooling by capturing and sinking water where it is wanted and keeping it out of areas where the excess water is unwanted. Sunken beds can be implemented as rain or bog gardens for growing food and ornamental plants that tolerate or need large amounts of moisture. Combined with other wetland strategies, sunken garden beds can be a great addition to a garden that may have too much water during certain seasons of the year.

Read more about Craig's Garden and view photos by following the link in Figure QR 3.14.

QR 3.14

Chapter 4

A Time For Everything: *Gardening Through the Changing Seasons*

"Spring is the time of plans and projects."
Leo Tolstoy, *Anna Karenina*

"It's spring fever. That is what the name of it is. And when you've got it, you want – oh, you don't quite know what it is you do want, but it just fairly makes your heart ache, you want it so!"
Mark Twain, *Tom Sawyer, Detective*, chapter 1

Residents of Phoenix march to the beat of a different drum, emerging from our climate-controlled homes and offices with a spring-like song in our hearts and busily planting when most gardeners across the nation are putting their plots to bed. The marvelous and peculiar reason for our offbeat behavior is simply that the invigoration of springtime comes twice annually to Phoenix; first at winter's thaw, and again when the sterilizing heat of summer finally abates. During both spring and fall, the desert buds and blooms, and its beauty is the just reward to desert dwellers who brave the torrid summers to live in the desert southwest.

I would dare to say that no one feels the passion and the promise of the changing seasons more than those who grow food in the arid low desert. Though our mild temperatures in the spring and fall are delicious, they are also fleeting. It often seems that just as soon as spring weather warms to conditions that are ideal for seed germination, temperatures quickly skyrocket to highs that are capable of damaging plant growth and production. In as early as February, 90°F days are possible, and sometime between Valentine's Day and Easter, I am often scrambling to apply shade cloth to protect my tender summer seed starts and to help my cool season vegetables and greens survive just a little bit longer.

By May, we begin to experience temperatures that in most other states would be considered *excessive heat*. The familiar archetype of the scorching desert is certainly confirmed by our record high of 122°F. Despite the heat for which we are famous, Phoenix is actually a region of high and low extremes, experiencing temperatures that dip as low as 17°F during the brief winter months. Though our lows pale in comparison to the sub-zero conditions experienced to the north, it definitely gets cold enough to damage a vegetable garden. Winters are especially hard on perennial plants that are adapted to the heat. These are relatively defenseless to frost, much like the humans, myself included, who live here specifically because we don't like cold weather.

Year-round gardening in Phoenix has certainly made me more attentive than ever of seasonal fluctuations. February 14th, Valentine's Day, is the date that we plant our tomato starts, and it marks the advent of the warm growing season. As the Fourth of July approaches, temperatures are soaring, and we begin to look forward to seasonal rains that signify the monsoon growing season. By Labor Day, Phoenicians begin to anticipate relief from the heat that is just around the corner, and we start to plant our cool season greens and vegetables that will thrive through the winter.

While we are blessed to have three distinct growing seasons, they are brief and overlapping. Diligence and constant vigilance are the hallmarks of successful valley growers who never rest, are always planning and planting and protecting their plots from the fluctuating elements. This is true for home gardeners and commercial growers alike, who become masters at season extension by necessity, employing various planting strategies and weather protection techniques in order to maximize productivity.

Success also depends highly upon precision in following a planting calendar that is tailored specifically for Maricopa County. General calendars are ill-suited to our climate, and the brief nature of our growing seasons requires that we plant seeds within short and specific date ranges. One of the mistakes that gardeners who are new to Phoenix often make is to plant seeds at the same times as they always did in other regions of the country. I was guilty of this error, planting tomatoes at Mother's Day, and wondering why fruit did not set. Once I learned to start slow-growing tomato seeds indoors in December and transplant them outdoors in February, production greatly improved. And when I added shade cloth to the tomato patch as the spring temperatures rose, I could barely keep up with the harvest. I share what I have learned about growing tomatoes at the link in Figure QR 4.1.

Phoenix is not alone in its seasonal challenges. All around the country, gardeners must adapt to local growing conditions: the fleeting summers of Alaska and Colorado, Florida's tropical heat and high winds, snow in the Midwest and the Eastern Seaboard. Gardeners in various regions have their own unique strategies to overcome seasonal impediments to growing and raising food. Local university extension offices and Master Gardener programs can be very helpful in sharing the various techniques that are employed in specific locations. Locate a Master Gardener program near you at www.ahs.org/gardening-resources/master-gardeners.

While learning seasonal techniques is helpful, understanding how the changing seasons affect plant germination and growth is even more important, as it will equip you with understanding that goes deeper than simple mechanics. In this chapter, we will explore how seasonal changes affect plants and livestock. We will look at the impacts that temperature and day length have on plant growth and germination, as well as how these factors drive egg production and livestock reproduction. Finally, we will learn how to make the most of the inevitable seasonal changes as they occur.

GROWING SEASONS

Vegetables require specific temperature ranges to produce crops, and are therefore classed as either warm- or cool-season, depending on the weather in which they best grow.

WARM-SEASON veggies require warm soil for seed germination and high daytime temperatures for steady growth and crop production. They are sensitive to cold weather, preferring temperatures in the 65–90°F range, and are easily killed by winter frosts. Included in this category are summer crops such as squash, melons, snap beans, corn, tomatoes, peppers, cucumbers, sunflowers and okra. You may notice that the edible part of most of these varieties is the fruit. These plants also tend to grow quite large, up and away from the earth. Figure 4.1 shows heat-loving Armenian cucumbers growing in our garden.

WARM SEASON

PLANTED IN SPRING & SUMMER

EDIBLE FRUITS

LIKE TOMATOES, PEPPERS, SQUASH, MELONS, CUCUMBERS, EGGPLANT, & CORN

It is interesting to note that although we plant bare root fruit trees in the winter when they are dormant, stone fruits and apples ripen during the warm season. Additionally, what we know as *winter squashes* such as acorn, butternut and banana are actually warm season crops that are harvested in the summertime. We call them *winter squashes* because of their long shelf life and their ability to last in cool storage without refrigeration for consumption through the winter.

COOL-SEASON veggies thrive at temperatures averaging 15 degrees cooler than those preferred by warm season varieties. Cruciferous vegetables, greens, root vegetables, peas and many varieties of herbs fall into this category. Most, including carrots, lettuces, Swiss chard and celery, are semi-hardy, able to endure light frosts. Others, such as kale, broccoli, cabbage and parsley, are hardy enough to tolerate hard frosts dipping into the 20s and teens. Figure 4.2 shows a large cabbage head forming in my winter garden.

4.2

COOL SEASON

PLANTED IN EARLY SPRING, LATE SUMMER OR FALL

HEADS, LEAVES & ROOTS

LIKE BROCCOLI, CAULIFLOWER, GREENS, CARROTS, RADISHES, LETTUCE, KALE & CABBAGE

Cool season crops are grown for their edible leaves, crowns, heads or roots. They tend to hug the earth or grow below the ground. They are best grown to maturity during cold or cool weather, started in late winter or early spring and again at late summer or early fall. They are not typically grown during the summertime; warm temperatures can produce a bitter taste or cause the plant to go to seed before the gardener has the chance to harvest the edible parts.

UNDERSTANDING *WHAT* AND *WHEN* TO PLANT

HARDINESS ZONES

Plant hardiness zones are guides that help gardeners to understand which plants will most likely survive at a given location. Knowing your hardiness zone is important. Seed packets often base planting recommendations on hardiness zones. Additionally, for plants that last year round, hardiness zones provide clues as to which will survive at local temperature extremes, particularly the average minimum temperatures of the zone.

The most well-known guide is the USDA Plant Hardiness Zone Map. Updated in 2012, the color-coded map is based on average minimum winter temperatures. Zones are divided into 10-degree Fahrenheit divisions, and further subdivided into 5-degree divisions. The higher the zone number, the warmer the climate. In Phoenix, it is recommended that we select plants that are hardy to zone 9b. Since zone 9b has an annual average minimum temperature of 25–30°F, plants that are hardy to this particular zone can withstand temperatures as low as 25°F.

If you happen to live in a microclimate that is a bit cooler or warmer than the surrounding climate, you may need to adjust which zone that you use as your guide. Tracking low temperatures in your garden for one or two years to calculate your specific average lows will allow you to compare your own data with the USDA data to determine if they match.

I often recommend using the zone that is one half-step up and 5 degrees cooler than your recommended zone, as temperatures can surprise us by dipping below average lows. Although Phoenix is zone 9b, with average temperatures as low as 25°F, I tend to select plants that are hardy to zone 9a, with average lows of 20°F. By doing so, when a frost hits, my garden is more likely to survive, even if temperatures happen to dip below our average lows.

// OVERLEAF // MAP 1: USDA Plant Hardiness Zone Map, 2012. Agricultural Research Service, U.S. Department of Agriculture. Accessed from http://planthardiness.ars.usda.gov

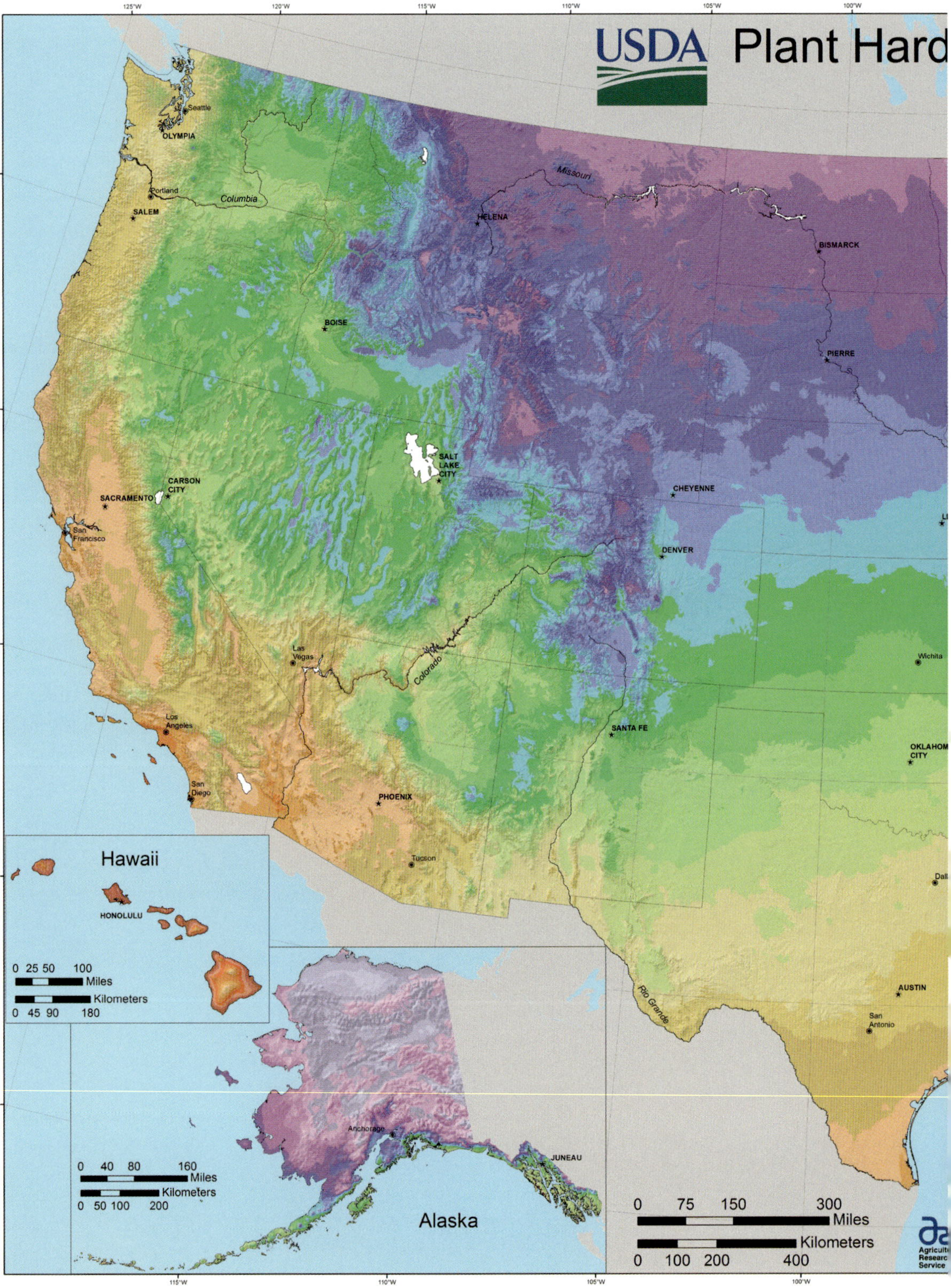

USDA Plant Hard

125°W 120°W 115°W 110°W 105°W 100°W

45°N 40°N 35°N 30°N 25°N

Seattle
OLYMPIA
Portland
SALEM
Columbia
HELENA
Missouri
BISMARCK
BOISE
PIERRE
SALT LAKE CITY
CARSON CITY
SACRAMENTO
San Francisco
CHEYENNE
DENVER
Las Vegas
Colorado
Wichita
Los Angeles
SANTA FE
OKLAHOMA CITY
San Diego
PHOENIX
Tucson
Dall
Rio Grande
AUSTIN
San Antonio

Hawaii
HONOLULU
0 25 50 100 Miles
Kilometers 0 45 90 180

Alaska
Anchorage
JUNEAU
0 40 80 160 Miles
Kilometers 0 50 100 200

0 75 150 300 Miles
Kilometers 0 100 200 400

Agricult Researc Service

115°W 110°W 105°W 100°W

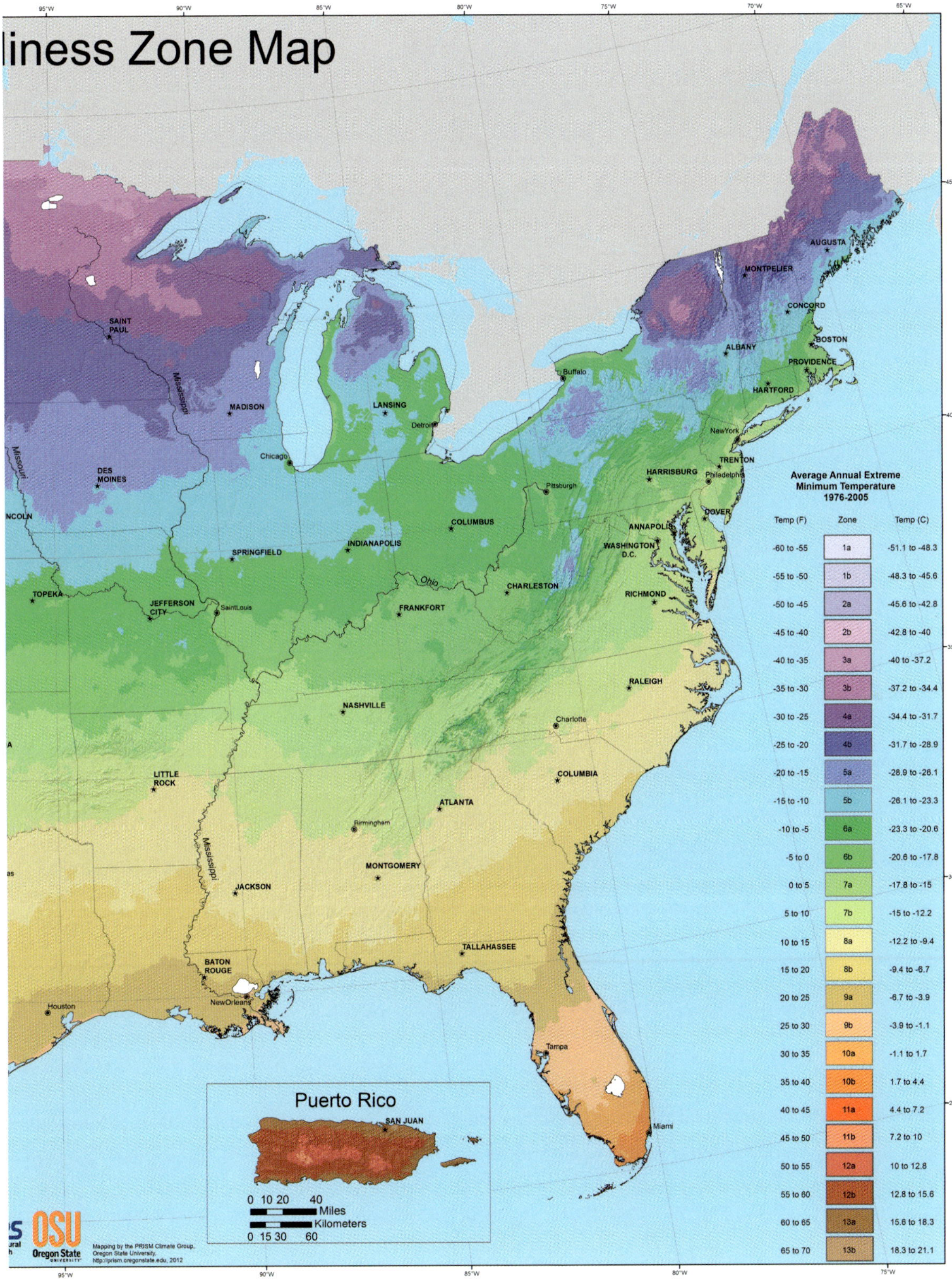

liness Zone Map

Average Annual Extreme Minimum Temperature 1976-2005

Temp (F)	Zone	Temp (C)
-60 to -55	1a	-51.1 to -48.3
-55 to -50	1b	-48.3 to -45.6
-50 to -45	2a	-45.6 to -42.8
-45 to -40	2b	-42.8 to -40
-40 to -35	3a	-40 to -37.2
-35 to -30	3b	-37.2 to -34.4
-30 to -25	4a	-34.4 to -31.7
-25 to -20	4b	-31.7 to -28.9
-20 to -15	5a	-28.9 to -26.1
-15 to -10	5b	-26.1 to -23.3
-10 to -5	6a	-23.3 to -20.6
-5 to 0	6b	-20.6 to -17.8
0 to 5	7a	-17.8 to -15
5 to 10	7b	-15 to -12.2
10 to 15	8a	-12.2 to -9.4
15 to 20	8b	-9.4 to -6.7
20 to 25	9a	-6.7 to -3.9
25 to 30	9b	-3.9 to -1.1
30 to 35	10a	-1.1 to 1.7
35 to 40	10b	1.7 to 4.4
40 to 45	11a	4.4 to 7.2
45 to 50	11b	7.2 to 10
50 to 55	12a	10 to 12.8
55 to 60	12b	12.8 to 15.6
60 to 65	13a	15.6 to 18.3
65 to 70	13b	18.3 to 21.1

Puerto Rico

0 10 20 40 Miles

0 15 30 60 Kilometers

OSU Oregon State University

Mapping by the PRISM Climate Group, Oregon State University, http://prism.oregonstate.edu, 2012

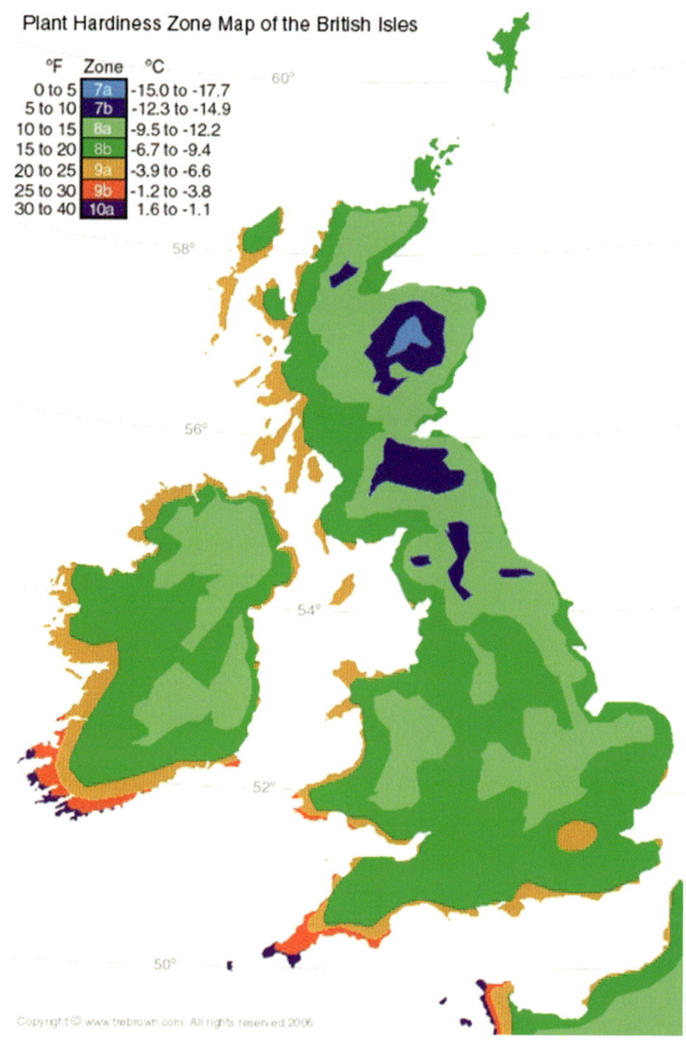

MAP 2: UK Plant Hardiness Zone Map. Credit: Trebrown Nurseries. Accessed from www.trebrown.com/hrdzone.html

PLANTING CALENDARS

If you are unsure about when to plant your garden, a planting calendar can be very useful. Although generalized planting calendars exist, since there is so much seasonal variation, I recommend that you get a planting calendar that is specific to your location or climate in order to know exactly what time of the year is most favorable for the varieties that you want to grow. Use the internet to find a calendar for your county, your USDA hardiness zone or your climate. As an example, for Phoenix,

I would search using keywords *planting calendar* and *Maricopa County*. I could also use the search terms *low desert* or *USDA hardiness zone 9b* or *arid southwest*. Another option is to contact your local Master Gardener program, many of which have their own localized planting calendar that they share with the public. By carefully following your planting calendar, you will increase your chances of success and minimize the expense and disappointment that comes from planting at the wrong time. Figure 4.3 shows an example of a planting calendar for low-desert climates that can be downloaded at UrbanFarm.org.

PLANTING CALENDARS

FOR VEGETABLES, FLOWERS & HERBS

TELL YOU THE OPTIMAL TIME TO PLANT IN YOUR SPECIFIC REGION OR CLIMATE

Low Desert Planting & Harvest Calendar

Brought to you by the Urban Farm, Matt Suhr and Greg Peterson

For information on classes and events offered on gardening and sustainability visit our website at www.urbanfarm.org and add yourself to our email newsletter or email Greg@urbanfarm.org or to contact Matt Suhr - happydirt@earthlink.net

KEY
- = Main harvest
- = Potential extended harvest season
- XX = Ideal planting time
- X = Good planting time
- • = Can be planted (with protective measure
- t = Set out transplants

Crop	Hardiness Temp	JAN 1	JAN 15	FEB 1	FEB 15	MAR 1	MAR 15	APR 1	APR 15	MAY 1	MAY 15	JUN 1	JUN 15	JUL 1	JUL 15	AUG 1	AUG 15	SEP 1	SEP 15	OCT 1	OCT 15	NOV 1	NOV 15	DEC 1	DEC 15	Comments
Artichoke-- Globe	20	•t	•t	t	t													X	X	xx	X	X	•t	•t	•t	Light frost helps first year harvest.
-- Jerusalem	<0	X	X	X	X	X	X	X																		Can be invasive.
Arugula	15	XX	XX	X	•													•	X	XX	XX	XX	XX	XX	XX	May be planted thickly.
Asparagus	<0			•	X	XX	XX	X	X	•									t	t	t	t				Don't harvest until 3rd year.
Basil	32	XX	XX	XX	XX	X t	X t	X t	X t	X	X	X	X	X	X	X	X							X	X	Start indoors in winter.
Bean--Blackeye	32					X	X	XX	XX	XX	XX	XX	XX	X	X	X	•									Performs well in full summer heat.
--Fava	20																		•	XX	XX	XX	X			Dislikes heat.
--Garbanzo	25	•	X	X	•														•	XX	XX	X	X	•	•	Stays low to ground.
--Green snap	32				•	XX	XX	•								X	XX	•								Seed will rot if planted in cold soil.
--Lentil	25	•	X	X	•														•	XX	XX	X	X	•	•	Harvest entire plant and thresh when dry.
--Lima	32				•	X	XX	X	•					•	X	X	X									Does best with trellis.
--Pinto	32				•	XX	XX	•	•						X	X	•									Harvest entire plant and thresh when dry.
--Soy	32				•	XX	XX	•	•						X	X	•									Use special varieties for edamame.
--Yardlong	32					X	X	XX	XX	XX	XX	XX	XX	X	X	X	•									Black-seeded types do best.
Beet	25	X	XX	XX	X	•											•	X	XX	XX	XX	X	•	•	•	Be sure to thin if you want big beets.
Bok Choy	22	•t	•t														•	X	XX	XX	XX	X t	X t	X t	X t	Bolts quickly in Spring.
Broccoli--head	25																X	XX	XX	X t	X t	•t	•t	t		Light frost improves flavor.
--Raab	25	X	XX	X													•	X	XX	XX	XX	X	X	X	X	Pick frequently to maintain production.
--Romanesco	27																•	XX	XX	X t	X t	•t	•t			Allow 15" spacing between plants.
Brussels Sprout	22															•	XX	XX	X	•t	t	t				Only early hybrids do well. "Oliver" is best.
Cabbage--Chinese	24																•	X	XX	XX	XX	X t	X t	•t		Heads form quickly. Be sure to thin.
--standard	26	t															•	XX	XX	XX	XX	X t	X t	•t	t	Red varieties take a little longer to head.
Carrot	23	X	XX	XX	X												•	X	XX	XX	X	X	X	X	X	Slow to sprout--mix in a few radish seeds.
Cauliflower	27																•	XX	XX	X t	X t	•t	•t			Fold leaves over exposed heads.
Celery	28															•	X	X	X	•t	t	t				Often stringy and bitter in desert conditions
Cilantro	28	X	XX	X	•													•	X	XX	XX	X	X	X	X	Flowers attract beneficial insects.
Collards	25	X	X	•													•	X	XX	XX	X	X	X	X	X	Light frost improves flavor.
Corn--flour	32				•	XX	XX	X	•				•	X	X	XX	X									Allow to totally dry on stalk.
--ornamental	32				•	XX	XX	X	•				•	X	X	XX	X									Plant in blocks for good pollination.
--popcorn	32				•	XX	XX	X	•				•	X	X	XX	X									Harder kernals than flour corn.
--sweet	32				•	XX	XX	X	•					•	X	XX	XX	•								Supersweet var. need very warm soil to sp
Cucumber--Armenian	32				•	XX	XX	X	X	X	X	X	X	X	X	X	•									Withstands heat better than standard types
--standard	32			•	•	XX	X	X	•						•	XX	X									Harvest frequently for best quality.
Dill	27	X	X	X	•														•	X	XX	X	X	X	X	Very easy from seed. Does not transplant
Eggplant	32	XX	XX	XX	X	X t	X t	X t	X	X	X	X	X	X	•									•	X	Best production in Fall.

4.3

FAST GROWERS

Perhaps because I garden in such a challenging environment, I prefer to grow fast-maturing vegetable varieties that are harvested in ninety days or fewer. On most seed packets and on some transplant tags, information is provided concerning the average number of days to maturity under optimal conditions. In a particular garden, the plant might mature faster or slower than average, depending on a variety of factors, including temperature, soil fertility, sunlight exposure and water availability. For example, if your soil is cool when you are planting seeds, it may take longer for seeds to germinate, which will in turn lengthen the days to harvest. The *days to maturity* information on the seed packet is useful to give you a general idea about how long it will take for your plants to produce their edible parts. But keeping a garden journal will give you a more accurate idea of harvest times in your unique garden.

When sowing seeds directly in the garden, *days* means the amount of time from seed germination to the first day that you can expect to harvest. If you are growing from transplants, then *days* generally means days from transplant to ripening. Figure 4.4 shows an example of how the days to harvest are displayed on a seed packet.

4.4

As I mentioned, my preference is to select plants that mature in ninety days or fewer. Recently, I was researching watermelon varieties using the Cornell University Vegetable Varieties for Gardeners database at vegvariety.cce.cornell.edu/. My family and my chickens love watermelon, and I like to grow three or four different varieties each summer. I came across two interesting melons: Sweet Dakota Rose with fifty-five days to harvest and Bush Snakeskin with 110 days to harvest. I would be apt to select

the faster growing variety which I could sow in April for a June harvest and again in July for a September harvest. Another perfectly acceptable strategy would be to plant both varieties at the same time in April, harvesting the Sweet Dakota Rose in June and the Bush Snakeskin later in August.

PLANT LIFE CYCLES

Plants are categorized by the length of their life cycles.

ANNUALS are plants that perform their entire life cycle in one growing season. Many garden vegetables are short-lived annuals that have to be replanted year after year. Beans, peas, broccoli, cabbage, dill, corn, squash and melons are a few common annual varieties.

PERENNIALS are plants that grow for many seasons. Often, they will die back in the winter and then spring (pun intended) back to life when temperatures warm. A few garden crops are perennial, including asparagus, rhubarb, artichoke, oregano, chives, mint, and many kinds of berries. Fruit and nut trees are also included in the list. Perennial food crops make gardening easy, coming back year after year with very little management required.

BIENNIAL plants require two years to complete their full biological cycle. The first season results in leaf and root growth, with flowering and seeding occurring during the second year. Beets, onions, carrots, parsley, Swiss chard and kale are examples of biennials. During the first season, we eat the leaves, roots or bulbs. Seed savers will leave a few of these plants to continue to grow into the second season, when a flower stalk appears. We call this *bolting*, and it is the time at which seeds are formed and can be collected for future planting.

Some perennial and biennial plants behave as annuals, or are treated as annuals by gardeners. Sage, peppers and tomatoes are examples of perennial plants that are often killed by frost in my garden. Although they could technically live year after year in their native regions, in my garden they act as annuals that must be replanted seasonally.

SEASONAL EFFECTS ON FARM ANIMALS

Plants are not the only ones on the farm whose fertility and growth are affected by the seasons. Poultry and game birds are particularly sensitive to day length, producing more eggs when the days are long and curbing ovulation during the short days of winter. Additionally, fertile eggs tend to hatch during the warm days of spring and summer when temperatures are right for embryo development.

The reproductive cycle of goats can also be seasonal.

ALPINE TYPE goats are seasonal breeders that have a defined breeding season, usually from late summer through December, with kidding occurring during the spring. The does will ovulate, or come into heat, about every twenty-one days. Bucks remain in rut during the entire breeding season. Most dairy breeds fall into this category.

EQUATORIAL TYPE goats originated in hot climates. These goats can ovulate at any time of year. Most of the meat breeds and pygmies are equatorial. We raised Nigerian Mini Goats at The Micro Farm Project, which were equatorial in origin but also wonderful for milking. What they lacked in quantity due to their small stature, they made up for in quality, producing sweet tasting milk with no *goaty* flavor. I typically bred my does in late fall for mid-spring births to coincide with our region's mildest temperature ranges.

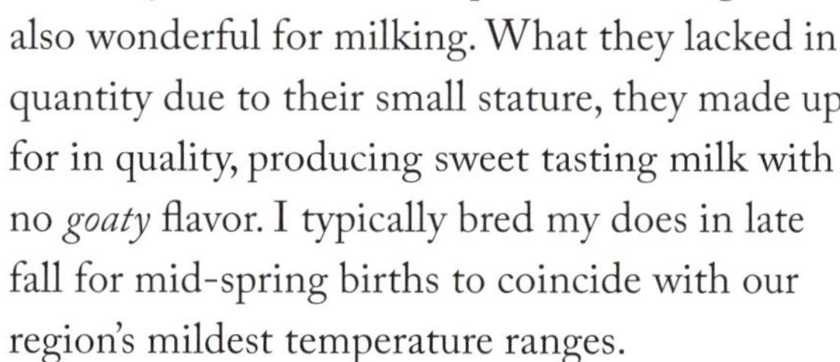

Our beautiful buck, Cassanova, provided breeding services year round to our own goats as well as to does from other farms. He was a very effective buck, kind to the ladies and popular due to his blue eyes and excellent lineage. He earned his keep year round in breeding fees. Although he was a great asset to our farm, I do not typically recommend that urban farmers keep their own bucks on site. Not only are they loud and possibly aggressive, they smell terrible when they are in rut. The smell can be offensive to neighbors, and if bucks are housed in close proximity with does, the scent can taint the flavor of their milk.

TURNING SEASONAL CHALLENGES INTO ASSETS

One thing that we can count on, that always stays the same, is *change*. Seasonal changes present their challenges to urban farmers, but they also provide positive benefits. Because vegetables grow at different temperatures, seasonal fluctuations offer variation to our diets. The end of the growing season is a time of rest, both for the soil and for the gardener. And seasonal weather changes help to keep pests and diseases in check.

But gardeners are not always satisfied to passively submit to the seasons, especially when we are enjoying a particular crop and want it to last just a little bit longer. At this writing, Arizonans are experiencing an unusually warm spring, and my beautiful purple lettuces are wilting in the heat long before they normally would. Thankfully, gardeners are not completely at the mercy of the weather. In fact, with the right tools and the right mindset, we can learn to extend the growing season while tapping into the benefits of cyclical weather patterns.

MARKET ADVANTAGES

When learning how to take advantage of seasonal changes, home gardeners can take a cue from market gardeners. Commercial farmers and gardeners are always looking for a niche product that they can bring to market. Growing and selling foods out of season is one way to create a profitable niche. In Arizona, farmers in Yuma grow lettuces during the relatively warm winters to bring to market when most other lettuce farms across the country are in their off-season.

In a very different climate from Arizona, Will Allen of GrowingPower.org produces kale and other vegetables during frigid Milwaukee winters using greenhouse structures called hoop houses. Check out Will's farm at youtube.com/watch?v=LRnulbOqo0k.

Whether or not your urban farm produces food for commercial sale, you can borrow techniques from commercial producers to create an economic advantage in your home garden. Market growers know that if they can grow and sell a particular product in the off-season, they will have little competition with other growers and can set their prices higher than would be possible during the regular growing season. As a home gardener, one of the ways that you can avoid paying inflated prices for produce at the supermarket in the off-season is to apply season extension techniques to grow vegetables and fruits slightly outside of the natural growing season in your area.

As an example, I take advantage of our mild winters and frost protection strategies to cultivate home-grown tomatoes in the off-season. By doing so, our family saves money and has delicious tomatoes when commercial tomatoes are expensive and relatively flavorless.

ROW COVERS AND FROST BLANKETS

Row covers are made from a garden fabric that serves the dual purpose of capturing warmth when temperatures are cold and preventing plants and soil from overheating when the weather is hot. Row covers are permeable, allowing light and moisture to penetrate. They can be placed directly on the soil or on top of plants, using mounded soil or garden staples to secure the fabric to the ground. As plants germinate and grow larger, row covers can be supported on hoops, secured with clips to prevent them from blowing away in the wind. As a side benefit, row covers also provide protection from damaging winds, pest insects and birds.

Frost blankets are a type of garden fabric that is similar to row covers, but tend to be made of heavier weight material. They are often used to protect landscape plants from frost, but are also useful in the vegetable garden in wintry weather. Frost blankets are applied when overnight freezing temperatures are predicted and are removed during the daytime when temperatures rise above freezing. If a hard frost is in the forecast, the cloth can be doubled up to provide greater insulation when needed.

Row covers and frost blankets must be draped all the way to the ground on all sides to prevent the cold from creeping in under the fabric. Use of row covers allows for early plantings in the spring and is also useful to keep cool season greens and vegetables growing longer into the warm season.

At The Micro Farm Project, I bend remesh supports into arches over my garden beds, sticking the ends down deep in the soil to secure them in place. Remesh is available at home improvement stores, typically near drywall or block fence supplies. Since we do not experience extended periods of freezing temperatures in Phoenix, I drape frost blankets over the arches and secure them only when a frost is predicted. If you live in a colder climate, you can leave the frost cloth on the supports all winter long. However, they do somewhat block airflow and sunlight. I recommend that gardeners vent or pull back the cloth during the daytime when temperatures warm above freezing to allow plants access to as much sunlight as possible and to release excess humidity.

In lieu of purchasing row covers or frost blankets, covering the garden with household blankets or sheets is a good alternative as long as the fabric remains dry. Wet fabric can lead to evaporative cooling, dropping the temperatures next to and under the blanket, which is the opposite of the desired effect. If sheets get wet, remove them during the day and dry them on a line or in a clothes dryer. Sunlight will recharge the heat stored in the soil, and the dry sheets can be returned to the garden in the evening to protect from nighttime and early morning frost. More information about garden fabric types and uses is located at the link in Figure QR 4.2.

WATER

It may seem counter-intuitive, but one strategy that commercial growers use to protect plants from cold temperatures is to water the soil prior to a frost. Water is an excellent insulator. Plant cells that are plump with water are more resistant to frost damage. Moist soil retains heat better than dry soil and conducts more warmth to the surface, warming the air above moist soil by as much as 5°F higher than the air above dry soil. Regular watering during periods of dry, cold weather can help protect plants from freezing temperatures. If your area experiences rainy winters and plants are at risk of over-watering, water thoroughly early in the day before a hard frost is predicted. It is not recommended to water if snow or ice is on the ground.

So, if your soil remains frozen all winter long, water everything well during the fall while temperatures remain above 40°F. More information about understanding frost and how to use water for frost protection can be found at the link in Figure QR 4.3.

GREENHOUSES AND COLD FRAMES

Greenhouses allow gardeners to extend the growing season for a few extra weeks and potentially year round with certain types of structures. But commercial greenhouses can be large and expensive, and not generally appropriate for most urban farms. Low-cost alternatives are available that meet the needs of the backyard farmers and gardeners.

One option is to buy a small backyard size greenhouse with clear plastic walls. I have found small greenhouses for sale at my local hardware store, larger nurseries, in seed catalogs and on the internet. At The Micro Farm Project, I prefer to make my own modified greenhouse using the remesh arches that were described previously. To create a small greenhouse, also known as *low tunnels* or *hoop houses*, drape the arches with clear plastic sheeting. During the daytime, the sun shines through the plastic, warming the soil. At night, the plastic cover traps the heat and keeps the garden bed warm. If a cold winter is predicted, I may place a blanket or length of frost cloth over the plastic at night. Or, I may string incandescent Christmas bulbs on the frames under the plastic and leave them on all night to provide some additional warmth. It is important to note that when using plastic, it should be lifted on one side for a few hours during the daytime to allow for airflow or excess humidity may build up under the cover. Select plastic sheeting that is strong enough not to rip easily. Wind and birds will damage thin, flimsy plastic, rendering it useless.

For seed starting, I use another strategy called a *box cold frame*. A box cold frame is a simple structure comprised of a box with a transparent lid that utilizes solar energy and insulation to create a warm microclimate within your garden. When my children found a 3×3ft window abandoned in our alleyway, I gave it new life as the lid of a cold frame. The frame is a wooden rectangle with the same dimensions as the window. The frame sits directly on the ground. The window serves as a lid, and it is hinged on one side for easy opening. The frame is located in a very sunny area on the south side of my property. Temperatures inside the structure are a few degrees warmer than in the surrounding garden. When starting seeds in trays, I place them in the frame and close the lid. The warmth and humidity that build up in the frame create the perfect environment for germination. For a few hours each morning, I prop the lid open to release excess humidity, closing it before noon so that it can warm up again before sunset. Tips and links for building and using cold frames are posted at the link in Figure QR 4.4.

SHADE CLOTH

We have covered quite a few techniques for frost protection, but what if you live in a climate that has very hot summers with temperatures rising above 100°F? In that case, your summer garden may benefit from shade protection in the hot afternoons. During the summertime, the angle of the sun is high and direct, creating intense light conditions that peak at noon. Temperatures tend to rise steadily into the late afternoon. The combination of intense sunlight and prolonged heat can stress garden plants, even those that are labeled as tolerant of *full sun* conditions. Plants may wilt in the afternoon heat, perking back up when the temperatures fall at dusk. Leaves and fruits may turn yellow or brown with sunburn. Tomatoes stew on the vine (this is *no* exaggeration in Phoenix.)

To give my garden relief from the summer sun and heat, I will use my remesh hoops to create a low tunnel (aka mini hoop house) to support shade cloth from late May through August. My preference is to use a loosely woven burlap or a 30% shade cloth. Both provide a little bit of shade protection without blocking so much sunlight that vegetable production is hindered. Burlap is relatively inexpensive, and shade cloth often goes on sale at the end of the warm season. I prefer to buy both in bulk by the yard, which tends to be more economical than buying a precut packaged length.

QR 4.5

Many garden centers carry shade cloth and burlap on bulk rolls so that customers can purchase as much or as little as they need. Pictures of remesh lattice, burlap shade covers and information about making low tunnels or hoop houses can be found at the link in Figure QR 4.5.

Unlike frost blankets and row covers, shade cloth does not need to be draped to the ground, unless it is serving a double purpose to prevent pests from accessing the garden. Many garden vegetables require pollinators to produce fruit, so you may need to lift the cloth during the early morning or at dusk when bees are active.

I mentioned in a previous chapter that I often drape shade cloth all the way to the ground in my tomato beds. This creates a cool microclimate under the cloth that maintains the right temperatures for flowering and pollination when the temperatures in the surrounding garden are too hot for either. The cloth also prevents hornworms and birds from attacking my tomato plants. I am often asked how my tomatoes are pollinated. A wonderful trait of tomato flowers is that they are self-fertile, requiring vibration to shake the pollen from the male part of the blossom to the female part. Bees are best, creating the perfect type of vibration to release the pollen. But, if I forget to lift the fabric to allow bee access, I will simply shake the tomatoes vigorously to release pollen. Some gardeners will touch a sonic toothbrush to blossoms to mimic bee vibration, and although it sounds absurd, it works!

CONTAINER GARDENING

Container gardens are a great option for growing sensitive plants or for growing plants in harsh climates due to the advantage of being mobile. I recently procured a small mango tree. Mangos are very susceptible to frost damage, so I have purchased a large plant container in which to grow it. I will place the container on wheels, which will allow me to roll the mango into a warm spot under cover of the patio during the winter, and return it to a sunny spot when there is no danger of frost.

Container gardens are useful in areas with short growing seasons. They can be placed in the sun when the days are long and warm, and moved inside to a sunny south-facing window or under a grow light to extend the season.

Containers may also be helpful in areas where high winds or hurricanes are common, as they can be moved into a protected area when necessary. For more information about container gardening, see Chapter 5 Section *INVENTING YOUR FARM*: Growing Food in Containers.

STRATEGIES FOR LIVESTOCK AND POULTRY

COOP LIGHTING

Chickens and other breeds of poultry, as well as Coturnix quail, can be stimulated to lay eggs during the short days of winter by adding incandescent lighting to the coop. Christmas lights will do the trick. A heat lamp with a white or clear bulb is also effective, with the additional benefit of providing warmth to the coop. We string Christmas lights inside our chicken coop and quail run, and hang a heat lamp out of reach of the birds. The lights are plugged into a timer that turns them on between 3 and 4am. They are scheduled to turn off at sunrise. This supplements the amount of natural light to which our hens are exposed by three to four hours. The goal is to expose the birds to fourteen hours of light each day to trigger egg production. By doing so, we continue to have fresh eggs all year long, though not as many during the winter as during the summertime.

As a reminder, I recommend having the lights turn on in the morning as opposed to the evening. The reason is that chickens have an instinct to roost at sundown. Lighting the coop may interfere with this instinct. When the lights go off suddenly sometime during the night, chickens that have not roosted may find themselves on the coop floor, unable to find their normal roosting space. This will cause the chickens stress and may hinder egg production.

Another reason that I provide lighting in the morning as opposed to the evening is to maintain our easy free ranging pattern. At The Micro Farm Project, I like to let my chickens out of their run to wander the property in the late afternoon. Though our backyard is larger than most urban yards, the chickens could cause significant damage to landscape and garden plantings if left unsupervised all day long. By letting them out an hour or two before sunset, the chickens have time to forage, and then they put themselves neatly away at dusk with no need for me to herd them back into the coop. My only job is to count heads to make sure that everyone is roosting, and then close and lock the door to the coop for the night. In the morning when the lights turn on, the chickens awaken, but they cannot exit the chicken run to explore the yard. By keeping them close to the coop, I ensure that they remain within range of the lights.

Supplemental lighting is by no means mandatory. Many urban farmers forgo the lights and allow the chickens to go out of production for the winter. The reason that I choose to keep my hens in production is that the lifespan of a chicken in Phoenix is relatively short due to our incredibly hot summers. Like humans, chickens are born with all of the eggs that they will ever have, and most do not live long enough to use them all up. We attempt to get the most production out of our hens by providing them with everything they need to live the longest, healthiest lives possible and by triggering year-round ovulation.

TIMED HUSBANDRY

I mentioned earlier in the chapter that we raised Nigerian Mini Goats and had our own buck, Cassanova. We also bred sheep. The Adam and Eve of our flock were Eleanor the Dorper ewe and Oliver the Barbados Blackbelly ram. Nigerian goats, Dorper sheep and Blackbelly sheep can be bred at any time of the year, so we were able to time kidding and lambing strategically. Since goat kids and sheep lambs are extremely sensitive to cold temperatures and do not regulate their own body temperatures well during the first few weeks of life, we bred our ewes and does in late-autumn for mid-spring births. In addition to cold temperatures, lambs and kids should not be exposed to wet weather. Since we had no completely enclosed shelters for our livestock, our aim was to have the kidding season occur mid-April through early May when temperatures are mild and typically dry.

Egg hatching can be strategically timed, as well. Hens do not usually sit on their eggs, also known as becoming *broody*, during the wintertime. However, with the assistance of an incubator you can time the hatching of eggs to meet your needs. I prefer to hatch eggs in the early spring, so that by the time the hens are five weeks old and ready to join the adults in the coop, temperatures are neither too hot nor too cold to cause the young birds stress. I will also hatch eggs at the request of customers who would like to buy chicks during the off-season when they are not available for purchase in our local feed stores. Figure 4.5 shows a few of our young birds housed in an outdoor pen in the springtime.

TIMING

BREEDING CAN BE SCHEDULED FOR CONVENIENCE AND PROFIT

BY CAREFULLY TIMING HATCHES, WE SAVED ELECTRICITY AND BROODED CHICKS OUTDOORS

4.5

If you are interested in hatching eggs, you will either need to have your own rooster to fertilize them or buy fertile hatching eggs. Roosters are not generally allowed in urban areas, but fertile eggs can be ordered from hatcheries online. You may also find a local source for fertile eggs, and I have discovered that a popular grocery store chain carries a brand of eggs that are generally used for culinary purposes, but that are also fertile and can be hatched. For more information about hatching, see Chapter 7 Section *Hatching*.

We have considered starting a business that would lease incubators filled with fertile eggs to local schools for science projects, collecting the chicks after the students have witnessed the hatch. We decided against doing so for practical reasons. Roughly 50% of the chicks would certainly be male, and we simply do not have the space to raise high numbers of roos to adulthood. Nor would we have the stomach to dispatch them in large numbers to the freezer. For someone with enough space and an outlet to sell the roos or the skill to efficiently process them, it could make for a profitable business model.

CONCLUSION

When it comes to the seasons, one thing that always remains the same is *change*. But with change comes the opportunity to find an advantage. By applying some thoughtful creativity, you can use the seasonal changes to create a market advantage or a personal advantage. So, although seasonal changes have their challenges, they also present opportunities for those who are willing to discover and seize them.

In the next chapter, we will explore space … Not the endless expanse of outer space, but the potentially limited space on an urban farm. How can a city farmer make the most of a metro property? There are many options for maximizing the space you have available, whether it's 10ft or 10 acres.

FEATURED FARM
WISH WE HAD ACRES

Lewis and I enjoy sharing our love for chickens with the public, so we volunteered our home to be featured on the 2012 Valley Permaculture Alliance Tour de Coops. Auspiciously, the pre-tour party for coop owners and volunteers was held at Wish We Had Acres Farm. The night of the party, I met Dr Dave and Laura, the proprietors of the farm. As I was leaving the party, Laura mentioned that she had a couple of Nigerian Mini Goat does for sale, just in case I knew anyone who was looking for goats to purchase. It just so happens that I had recently become convinced that I wanted goats. I asked if I could come back again to see the does in the daylight and learn more about them. I returned the following week and purchased two lovely goats, Annie Oakley and Phoenix Rose. Thus, a friendship began, based on a mutual interest in urban farming and love for dairy goats.

Dr Dave and Laura helped us out in so many ways, from teaching us how to keep goats, to giving vaccinations and disbudding goat kids, and even rescuing us with emergency milk when our lamb Eleanor arrived at the farm as an orphan. We were very sad in 2014 when Wish We Had Acres pulled up roots and moved to one of the last standing farmsteads in Charlotte, where the farm expanded from ¼ acre to 5 acres. Of course, we were happy that they had achieved their dream of having acres on which to farm, and promised to keep in touch.

Not long after the move, Laura posted a lovely picture of their barn surrounded by a dusting of snow. Brrrr! What an adjustment it must have been for both humans and animals to move from sunny, warm Arizona to a snowy winter in North Carolina. Laura posted adorable photos of goats and goat kids in jackets and sweaters. Figure 4.6 is shown here courtesy of Laura Denyes and Wish We Had Acres.

I asked Laura how they coped and for a few tips concerning cold-weather farming. "Honestly, goats develop an undercoat of cashmere so the best plan is to keep their hooves trimmed for wet weather and keep their housing well insulated." She also mentioned that dry shelter is key for everyone. To keep

4.6

housing dry, make sure that you balance insulation with ventilation, and add dry bedding to your stalls and coops regularly.

Wish We Had Acres uses a combination of heated buckets and warm water to prevent water buckets from freezing. "For the animals without heat-regulated buckets, we haul out hot water sweetened with molasses." Molasses provides a quickly source of energy for goats, which is so important in cold weather when animals are expending lots of energy keeping warm. It also serves as a source of minerals that may be missing from their diet when fresh, green pasture is not available and supports good fur growth. Laura cautions, "Diet is everything in colder weather." Dry, mold-free hay and minerals will keep them healthy. Some goat experts also recommend supplemental protein during the wintertime.

Laura recounts an incident that occurred shortly after their move to NC. "Our first winter here from AZ, the goats hadn't had a chance to develop those lovely undercoats … so we headed to Goodwill and promptly dressed them all in hoodies and sweater vests. Doncha know, they found a hole in the fence and marched across the road in their get-ups. It was a hoot! Traffic stopped five cars deep in both directions. Drivers were all too amused to be upset. Welcome to the neighborhood!"

Read more about Wish We Had Acres and see photos of goats in coats at the link in Figure QR 4.6.

THE DIRT on farming

Wish We Had Acres Full Interview

CITYFARMINGBOOK.COM/WISH-WE-HAD-ACRES/

QR 4.6

Chapter

5

Make Room For Abundance:

Utilizing Urban Spaces Efficiently to Grow and Do More

It was the second to last day of 2012. The excitement and busyness of Christmas was over, and we were beginning to shift gears into plans and dreams for the New Year. The phone rang, and the voice on the other end of the line was Mary, a friend we knew from church. Little did I know that this call would change the course of 2013 for our family and for our farm.

Mary's young daughter had received a peculiar Christmas gift, adorable, furry, with four legs and a wagging tail. But it was not a puppy. The gift was a black and white ewe lamb, no more than a day or two old, whom her daughter had named Eleanor. To understand the oddity of the gift, one must know that Mary is not a farmsy-type person; she does not own a large plot of property and has no experience with livestock of any kind. Why her family would receive such a gift was a mystery. Fortunately, I was first on her list of people to call for help.

The tiny lamb would not take a bottle, and was weakening. Mary asked if I knew anyone who might want to adopt her, and could we keep her at our farm until a new home was found. Lewis and I immediately grabbed a blanket and drove across the valley to the rescue.

On the way out of Mary's door with the lamb in my arms, I had assured her daughter that we would find a good home for Eleanor. Riding home with her wrapped in a blanket on my lap, it became clear that her new home would be ours. I had fallen in love. Figure 5.1 shows a photo of that fateful ride.

5.1

New Year's Eve was spent nursing Eleanor back to health and bonding. Attempts to coax her into taking a bottle were unsuccessful. Finally, out of desperation, I filled a bowl with water and set it in front of her. Eleanor stuck out her tiny lips and sucked up some liquid with a loud slurping sound. Relieved and elated, I replaced the water with raw goat milk, which she drank from a bowl until she was weaned. Even as a grown ewe, Eleanor continued to drink loudly, slurping water from her dish just as she had as a tiny lamb.

Though I had never allowed animals in the house, Eleanor lived inside and was free to roam. She often disappeared, and we would find her curled up in a closet sleeping in the dirty clothes pile. When she was awake, she was in perpetual motion. Like the children's song, *Mary Had a Little Lamb*, she followed me everywhere I went, tiny hooves clicking on the laminate floor behind me. As a flock animal, she would not be satisfied with my company for long. She needed another of her own kind. An attempt to introduce Eleanor to our goats was unsuccessful. Annie Oakley and Phoenix Rose were rude hostesses who greeted Eleanor by rearing up and butting the tiny lamb with all of their strength. For her safety, I swiftly brought Eleanor back into the house and pondered a new plan.

A second ewe lamb would have been a suitable companion if I had not begun to dream of a small flock of sheep peacefully grazing the lawn to perfection, of raw sheep milk for artisan cheese, and of home-grown racks of lamb for the holidays. Scanning the internet, I found a Barbados Blackbelly ram lamb for sale in our area. We fell in love with him as we had with Eleanor, and the children named him Oliver.

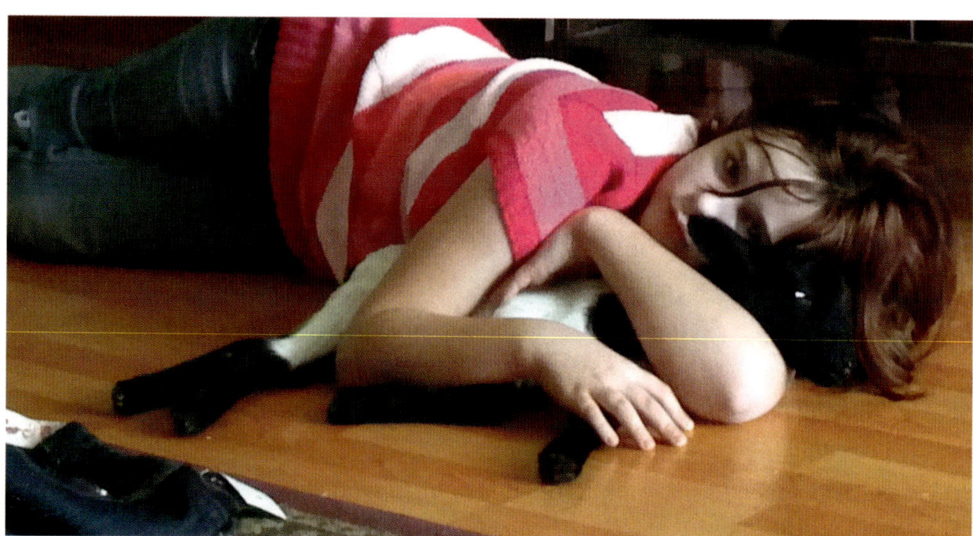

Oliver and Eleanor bonded immediately, although Eleanor, who had initially refused to take a bottle, became jealous of Oliver's. Rudely butting in on his feedings, she was relentless in attempting to steal his milk. Fortunately, they weaned quickly and were able to move outdoors to a 5ft dog kennel. But bigger problems quickly surfaced.

In my zeal to start our little flock, I had failed to educate myself properly on sheep. I learned in the nick of time that sheep can potentially become fertile before five months of age and should be housed separately to prevent a dangerously early pregnancy. We scrambled to purchase and build a second chain link kennel so that the lambs were technically separated, but could see each other. At night, the two would sleep back to back, each in their own pen, touching fur through the fence.

Meanwhile, housing challenges were rising on other fronts. Goats were dwelling in the chicken run with our hens, but their persistent attempts to break into the poultry feed convinced me that they needed their own space. Kind friends and family members converged on our property to help us fashion a goat habitat out of recycled materials, complete with two trellises bungeed together to serve as a gate.

In addition to the four-legged animals, birds of various sizes squawked and chirped in their own spaces; chickens in the girls' playhouse, turkeys in a repurposed rabbit hutch, and quail in the aviary built by a skilled friend.

The backyard was beginning to look like a makeshift zoo. Slapdash animal enclosures were strewn haphazardly across the yard, with no rhyme, reason or discernable plan. The arrangement was neither appealing to the eyes nor easy to manage. Our morning routine became arduous. Rising with the sun, the family hurried to feed hungry, noisy animals, dragging water hoses across the lawn through a maze of pens, toting feed and tossing hay.

Free time and extra money waved goodbye, consumed by repairs and adjustments to enclosures, cleaning up animal waste, purchasing and hauling feed, and so forth,

in an endless parade of tasks and expenses. Problems arose too often, and solving them with stopgap measures became our habit. One of our most time-consuming challenges was providing protection from the elements. When we assembled the dog kennels for the sheep, we initially failed to roof them, thinking that we had plenty of time to come up with some sort of covering as it rarely rains in Phoenix. But, of course, as the clouds were rolling in, I was rushing around the backyard securing tarps and old political signs to the top rails of the enclosures to provide shelter in the fastest (and *ugliest*) manner possible.

We told ourselves that it was only *temporary*. But I have learned over the years that our family practices a form of collective denial, too often allowing the *temporary* to become permanent by lack of attention. Examples abound: the cheap mini blinds that have hung temporarily in the kitchen for seventeen years; unirrigated vegetables, temporarily hand-watered for too many seasons; a pile of mulch blocking the RV gate for six months. These have become part of the scenery of the farm. The thought that we should come up with more solid solutions occasionally alights like a butterfly in our minds, swiftly flying away on the breeze, forgotten.

And so, the temporary kennels with the tarps persisted for three years, housing Eleanor and her offspring until the last of the sheep had left the property. Due to his size, Oliver had graduated to a steel enclosure. Amazingly, he figured out how to open the complicated gate latch, and I had to rescue Emily, who was protecting herself with a fence post driver as he reared back to charge her. Despite his instinct to charge and butt, Oliver craved attention and loved to be scratched over the fence. It became difficult to clean his pen as he would follow me around, nudging me with his muzzle to request a scratch. Unlike flighty Eleanor, he would stand stock still to be brushed and would demand my attention if I was anywhere within eyesight.

In 2015, the lambs had been processed and Eleanor had been put down for a medical reason. My buddy Oliver was the sole sheep on the property. He was lonely, and it became clear that we needed to find a new home for him. Always the ladies' man, we were fortunate to find a ranch with dozens of ewes that had recently lost their only stud. Oliver stepped into his shoes, and by all reports is a very happy ram.

Writing about our adventures with the sheep and their offspring brings up feelings of nostalgia and mild regret. You can read more about our early adventures following the link in Figure QR 5.1, which tells the full story of our animals. I miss their presence on our property, especially of Oliver whose quirky personality brightened my

days. So why did we decide to rehome him after Eleanor passed rather than start a new flock? The decision boiled down to space, or the lack thereof.

Although we live on a large enough city lot to legally keep sheep, it had become very challenging to do so. There simply was not enough space both to house the sheep and to grow food for them to eat. Truckloads of purchased alfalfa and Bermuda bales drained the farm budget. To supplement their diet and reduce the feed bill, I propagated Morning trees and fodder that grew speedily. We fed them scraps from our own garden and from a nearby school garden. They dined on palm fronds, corn stalks and weeds. But we simply could not grow or scavenge enough supplemental food to support the sheep as well as the goats and their various offspring.

Despite copious amounts of food given to them each day, every bush or tree within reach was neatly trimmed by their nonstop nibbling. And although sheep don't eat grass to the ground, what they failed to nibble they trampled to oblivion. Pens were raked clean twice weekly, but during the days in between, their hooves pressed manure and uneaten alfalfa into dry layers that smothered the lawn. Every place that they roamed turned to bare, brown dirt. This was truly an amazing turn of events considering that our lawn had previously been covered by aggressive and invasive Bermuda grass, the bane of Arizona gardeners who battle tooth and nail to keep it out of their vegetable beds. I have found the secret to easy Bermuda remediation: get sheep. Under their constant foraging and stomping, the lawn rapidly disappeared.

We simply had too many animals in too small a space at too high a cost. I purposefully avoided calculating the cost per pound of the lamb that ended up in the freezer, but I am certain that it would be less expensive and more sustainable to purchase pastured lamb from a local farmer than to continue raising sheep in my backyard.

When designing an urban farm, space is always a consideration. The question arises, *what do I have room to grow or raise*? Square footage is important, but the weightier question is, *what do I have room to grow or raise sustainably*? In other words, will it meet the following criteria?

QR 5.1

- Will it be good for the environment?
- Will the return outweigh the financial cost?
- Will the return outweigh the labor?
- Finally, can it become part of a *regenerative* system?

A *regenerative* system is self-sustaining and gives back more than it takes. Clearly, our system of raising sheep had been *degenerative*, destroying the lawn, depleting our finances, and wearing us out. This was the opposite outcome from what I had envisioned and attempted to create. So, now that I have explained what went wrong, how could we have designed the system differently to make it *regenerative*?

This chapter will answer that question and explore the concept of regenerative design. We will discover the many ways that city dwellers can maximize limited urban space to grow and raise as much healthy food as possible, sustainably, using regenerative systems that improve our homes, our health and our communities.

HOW DOES NATURE DO IT?

To begin our study of regenerative design, I want to share an epiphany concerning the old story of the Garden of Eden that changed how I view my own garden and, by extension, my urban farm. Here is my paraphrase of the story, which is not at all theological, but strictly a gardener's perspective. According to the account in the Book of Genesis, God created a perfect world. The first humans, Adam and Eve, were placed on that perfect world in a beautiful garden, and were charged as caretakers to the plants and animals therein. The couple enjoyed the abundance of the garden, lacking nothing. They were completely free, naming the animals and tending the garden as they saw fit. And though it came with the warning of eventual death, they were free even to eat the fruit from the prohibited Tree of the Knowledge of Good and Evil. Lured by the serpent's promise that if she ate the fruit, she would gain God-like wisdom, Eve succumbed to the temptation. Adam quickly followed suit. As a result, the couple were cast out of the garden, doomed to scratch a meager existence out of the ground by the sweat of their brow.

When I became intrigued with Permaculture, this story leaped off the pages as a fitting analogy for regenerative *vs* degenerative food production. At first, Adam and Eve grew food in a regenerative natural ecosystem that was abundant and self-sustaining.

When I think of Eden, my imagination conjures up the tropical rainforest of today. Powered by sunshine and natural rainfall, the forest provides its own food sources via complex food webs. Waste products are decomposed and recycled; pests and diseases are controlled by predators and weather patterns; new plants and animals replace the dead. A resilient ecosystem can sustain and regenerate itself for hundreds or even thousands of years.

In contrast, my own garden resembled nothing like an intact, regenerative ecosystem. Although I gardened in part to *get in touch with nature*, there was not much that was natural about producing food using machines, chemical fertilizers, pesticides and a tremendous amount of labor. In fact, I spent a lot of time fighting off nature – pulling weeds, killing pests, disposing of debris and keeping everything tidy in neatly organized boxes and rows. The system was certainly not regenerative. Instead, it was *extractive*, depleting the soil and emptying my pocketbook. It was also clearly *degenerative*, illustrated by the condition that even when things were going smoothly in the garden, happily growing plants would wither and die without constant management and life support in the form of irrigation and fertilizer.

This was the only way that I knew how to garden. Slowly, however, I began to realize that I had a choice. I could grow food by the sweat of my brow, like Adam and Eve after the fall. Or, I could choose to return to Eden. Not literally, of course, but in the sense that I could tap into the natural systems and rhythms to produce abundant food in a regenerative system.

I began to read about food production systems that mimic nature, and really wanted to give it a try. It was a little bit overwhelming. Natural systems that are highly resilient are also large and incredibly intricate, so complex that modern science barely understands all of the processes and relationships that keep these ecosystems running. But the main point to remember about regenerative systems is that they supply their own needs in what's called a *closed loop* system. In a perfect closed loop, no *inputs* need to be brought in to the system to sustain it, and no waste products are created for which there is no use. The earth itself is a closed system, meeting its own needs in house, recycling waste products, and regenerating itself constantly. It only has one input: sunlight. The earth is subdivided into large biomes and smaller ecosystems that meet most of their own needs, but require more outside inputs, such as wind, sunlight and rainfall. In nature, the smaller the system, the less resilient it becomes and the more outside inputs it requires. [The Encyclopedia of Earth (2014) Ecosystems. Available at: http://editors.eol.org/eoearth/wiki/Ecosystems. (accessed November 2016).]

Human ingenuity adds a whole new dimension to the process, making regenerative design possible even in small-scale systems. Although my urban farm would likely never become an entirely self-sufficient system, I began to look for ways to close as many loops as possible to create small scale, intensive food production systems in my backyard. The aim was not only to be ecologically sound, but also to maximize productivity on my small property, while minimizing expenses of time and money. Whew! That was a tall order!

I had to keep it simple, so I started connecting just a few elements of my farm to create rudimentary closed systems. Closed loops are all about relationships amongst the elements in an ecosystem. One element's waste is another's food; one's behavioral patterns assist the greater community; one's biological characteristics serve a higher purpose. In light of the big picture, I began to ask the question, *what can the elements of my farm do for each other?*

The first closed loop system that I created included just three elements: compost, a garden, and chickens. I noticed that food scraps in my compost pile attracted some bugs. Chickens love to eat bugs, as well as food scraps. They also love to dig and scratch, which resembled the work that I was doing to turn my compost piles. Perhaps I could throw my compost materials into the chicken run instead of the bin that I was currently using.

The more I thought about the idea, the better it seemed. Chicken manure is a fantastic fertilizer once it is composted, so why not let the bird's poop directly into the compost pile rather than adding it to the pile later? And the chickens were great at spreading out bedding in their coop, so why not let them spread compost in the garden?

And so, I began collecting compostable materials in the chicken run behind the coop. Kitchen scraps, landscape trimmings, weeds, garden clippings, and waste from the other animal pens were all thrown into the pile. The chickens gleefully scratched through the pile, searching for tasty tidbits. Figure 5.2 shows the hens pulling scraps out of a compost bin. Each day when I refreshed the drinking buckets, I discarded the old water on the pile to keep it moist. This renewed the chickens' interest in the pile, which they energetically shredded into small pieces that decomposed rapidly into marvelous compost. My feed bill decreased. I have read that chickens who consume a varied diet produce healthier eggs, so I can only assume that our hens' eggs were not only delicious, but more nutritious, too.

CLOSED SYSTEMS

CREATE SYSTEMS THAT MIMIC NATURE AND CONNECT FARM ELEMENTS

MAXIMIZES PRODUCTIVITY

MINIMIZES WORK, EXPENSE AND INPUTS

5.2

I began to compost more materials than I had previously when I was using a bin composting system. The old system had not broken down waste from my animal pens quickly enough to avoid bad odors and flies, so I had resorted to throwing a lot of it away. I had also avoided putting anything with meat or cheese in my compost bin for the same reasons. But, with the new system, the chickens were able to break down both the pen waste and the meat scraps quickly enough that flies and odors disappeared.

Within a few short months, I had enough compost for my farm and plenty to share with a community garden. I noticed a marked improvement in the quality of my soil, and consequently in the health and productivity of my plants. Less and less fertilizer was required to grow more and more plants. Higher yields equated to more food for my family and more scraps to feed the animals, which reduced my grocery and livestock feed budgets. Increased organic matter improved the soil's water holding capacity, and I joyfully discovered that my water bill plummeted.

The chickens seemed extremely happy and healthy. I, too, was happy to be spending less time tending the compost pile and less money on food, water and fertilizer. And the community garden was delighted to receive free compost.

I am now looking at ways to refine the system and close more loops. The biggest open loop is chicken feed that I purchase from a local feed store. Perhaps I could use the compost to cultivate native, drought-tolerant grains to supplement their

QR 5.2

diet. I am also considering dragging my chicken tractor into the alley and letting my hens dine on the weeds that grow there. Additionally, I am sectioning off areas of the farm and planting more edible trees so that the hens can free range longer each day and have more to eat in the yard. More details about my thought process concerning connecting elements and resources to deepen understanding of the concept are posted at the link in Figure QR 5.2.

BIG YIELDS IN SMALL SPACES

We have seen how systems that are designed to mimic nature benefit a farm. But, how does a discussion of closed loop systems specifically relate to the topic of *space*? In urban environments, land for growing food can be a scarce resource. When we find a little spot of ground on which to grow, we often want to maximize its potential. By thinking in terms of creating closed loops, we can begin to produce as much food as possible in our limited space, while conserving other limited resources, as well. As we explore a few of the many techniques that growers use to produce food in urban spaces, let us keep in mind the big picture of regenerative design. In so doing, we can apply our creativity to combine the techniques in novel ways and fashion our own unique systems that are increasingly productive to feed our families and our communities.

INVENTING YOUR FARM

One of the key pieces of creating a regenerative farm design is simply deciding what elements to include on the farm and where to place them. The following steps will make the process easy:

STEP 1: TAKE INVENTORY OF WHAT YOU ALREADY HAVE

If you carried out the steps to begin the farm planning process in Chapter One, you have already begun to observe your property to assess what you already have. Perhaps you took some notes on your observations or drew a map. If you have not already done so, observe your property and gather information on the various sectors, or *microclimates*, such as the sunny and shaded areas, windy locations, wet or dry locations, particularly hot or cold spots, and slopes.

Make note also of the structures on your property. Do you have sheds or other outbuildings? Walls or fences? Existing gardens and other landscape features? Materials that you could use for the construction of new elements? And are there any features that you could move or remove? Make a list of these, or draw them on your property map. Figure 5.3 shows what my list may have looked like.

5.3

TAKE INVENTORY

CONNECT THE DOTS; HOW CAN THE ELEMENTS SUPPORT EACH OTHER?

CONSIDER THE VALUE OF EACH ELEMENT, ITS CHARACTERISTICS, BENEFITS AND HOW IT COULD BE EMPLOYED ON YOUR FARM

STEP 2: ANALYZE POTENTIAL NEW ELEMENTS TO ADD TO YOUR FARM

The word *analyze* sounds technical, but analyzing elements is actually a very simple process. For any item that you want to add to your farm, make three brainstorming lists. The first list will be *what the element needs* and the second will be *what the element can do*. For the second list, think outside the box. Include the benefits of the element, what the element produces and its characteristics. Don't forget to include negative characteristics, too. Sometimes the drawbacks of one element can be solved by another, or even be of benefit to another. If I were analyzing corn, chickens and goats, I might list the products as ears of corn, eggs and milk. Add to the list how you could use other parts of the element, such as feeding cornstalks to livestock, using feathers for crafts or feather meal, collecting goat manure for composting. Jot down useful characteristics and behaviors as well.

The third list is used to envision possible connections between elements to create mutually beneficial relationships. Perhaps you could grow beans with the corn; beans would use the stalk as a trellis and the corn's need for nitrogen would be supplied by the beans. Maybe you could grow watermelons in the corn patch between the stalks to provide ground cover to the soil and food for humans and chickens. At the end of the season, goats and chickens could be allowed to access the area; the animals would receive supplemental food and exercise, and the bed would be prepared and fertilized for the next growing season. When it comes to the potential productivity of the systems that you create, the sky's the limit! So take some time to *ruminate* on the possibilities.

STEP 3: DECIDE WHERE TO LOCATE ELEMENTS ON YOUR PROPERTY

The next step is to draw out a map of your property with potential locations for each of the elements. It may be helpful to draw the map on graph paper to scale, or at least somewhat to scale. I am personally not very good at detailed measurements, but having a drawing that is at least close to scale is very helpful. Draw pictures of your elements also to scale on your graph paper. This could just be circles or boxes, or it could be more artful. The important thing is to represent the approximate square footage of space that each element will require. Move the drawings around on your property map until they fit together in a way that is functional and pleasing to you.

Here are some considerations when placing elements in the design:

- Placement should be legal. For example, in Phoenix, chicken coops must be located at least 80ft from the nearest neighbor's house. That limits where I can place a coop on my property, and there may be similar rules in your municipality.
- Place elements that you need to access daily close to the house. I collect eggs and herbs daily, so those elements are located near my back door. Vegetable gardens and fruit trees are a little bit farther out, and the wood chip pile that I only occasionally visit is at the very back of the property.
- Place elements in advantageous microclimates. At the Micro Farm Project, water harvesting barrels and the chicken coop are located on the north side of the house, where they receive more shade and stay cooler in our hot climate. Vegetable gardens are on the south and east sides of the property where they will receive the most sunlight.
- Place elements where they won't become a nuisance. This is especially pertinent to large plants, such as bushes and trees, whose growth patterns can damage structures, block views and shade out vegetable plantings. It is a hassle to continually trim or prune large plants that are placed poorly, and they are not easy to move once established.
- Place elements for which you want to create connections in close proximity. A compost pile can be conveniently located next to a garden. If chickens will be allowed to access goat pens for waste cleanup and pest control, place a gate between the coop and the pens. If you want your kids involved in gardening, locate the vegetables near their play areas.
- Observe and adjust. The previous steps will help you to design an urban farm that is highly productive and easy to manage with as few problems as possible. If unforeseen issues arise or implementation reveals flaws in the system, adjustments can be made along the way. Farms tend to be endless *works in progress*. This is certainly true at The Micro Farm Project. Changes occur often as we gain new information and learn from our mistakes.

If you already have a piece of property on which to farm, you can begin these steps right away. If you are still looking for a location on which to farm, it may be helpful to begin brainstorming which elements you might want to include on your future farm, and the connections that you could possibly make. This will be good practice and will make it easier to develop a plan once you have procured a location.

TURNING URBAN CHALLENGES INTO ASSETS

Finding land for growing food in the city can be challenging, but innovative farmers and gardeners are applying creativity to put unused and underutilized areas into food production. If you are looking for land or want to expand your current space, here are some ideas.

EMPTY, ABANDONED AND UNSIGHTLY LOTS

The financial crisis in 2007 and 2008 took a toll on the commercial real estate market in Phoenix. Thousands of acres of land that had been slated for commercial development sat vacant and unused. To improve the situation, developers and urban farmers joined forces to turn these lots into urban farms. Farmers were now able to grow food on inexpensive or free land. Land owners benefited from the reduced tax bill associated with designating properties for agricultural use. And neighborhoods were improved by fresh food, beauty and community relationships that grew on the farms. One of the most well-known farms to lease underutilized land is Crooked Sky Farms, owned by innovative farmer Frank Martin. Frank leases several parcels in Phoenix, in industrial areas and near an interstate. You can read more about the farm at CrookedSkyFarms.com.

In cities all across the United States, lots that are abandoned for one reason or another can become an eyesore and bring property values down. In response to blight caused by abandoned or neglected sites, some gardeners have resorted to guerrilla gardening tactics, gardening on land that the gardeners do not have the legal rights to utilize.

Some guerrilla gardeners carry out their actions in secrecy, to make an area more productive or attractive. Others garden visibly to attract publicity and generate change. In response, many cities have initiated programs by which gardeners and farmers can access vacant lots for agricultural use or community managed open spaces. One example is Baltimore's *Power in Dirt Initiative* that allows residents to adopt vacant lots for gardening or greening purposes. You can read more about this program and others like it at UrbanAgLaw.org/land-access.

FARMING GETS OFF THE GROUND

Densely populated cities have limited space for growing food at ground level, but acres upon acres of wide open space in some unusual places. That is why, instead of competing with developers for premium ground space, farming operations are moving up in the world – up as high as the rooftops. On a recent visit to New York, I was astounded by the number of rooftop farms in the five boroughs. Some of the farms, such as the Brooklyn Grange and Gotham Greens, are large-scale operations providing copious amounts of produce for the area's numerous grocery stores and restaurants. Although these farms are truly amazing, they are not easily replicable. I was lucky enough to tour three smaller-scale farms that each demonstrated growing methods that anyone with a suitable rooftop or balcony could copy.

The first was Rosemary's restaurant, located in the West Village of Manhattan. Chef Wade took me on a tour of the garden, which is located one flight up from the restaurant and is open to the public. Rich soil and rows of greens and vegetables give the garden the feel of a traditional at-grade garden until you look up and see at eye level the tops of the trees growing in the garden across the street. With help from the Brooklyn Grange, the roof was covered with a waterproof membrane and a thick layer of felt that acts like a giant sponge, holding moisture in an environment that would otherwise dry out rapidly. Thousands of pounds of soil were transported to the roof with the aid of a crane. The garden grows greens and vegetables that serve the kitchen below. Although this garden is much larger in scale than most individuals would need or could afford to assemble, it is essentially an oversized raised bed. This illustrates that raised bed gardens are a very viable option for rooftop farming as long as measures are taken to conserve water and to prevent runoff from damaging the roof. To see what is happening now in the garden, take a look at the live rooftop webcam at http://rosemarysnyc.com/about/rooftop-webcam.

The Hell's Kitchen Farm Project takes a much different approach to water conservation and management. The 4,000 sq ft roof at the top of Metro Baptist Church is dotted by plastic kiddie pools that serve as raised bed gardens. Each garden consists of a small pool drilled with drainage holes located at the lower edge of the pool wall. Pools with drainage holes are nested inside a second, intact pool. This approach allows moisture to drain from the soil in the upper pool into the empty space in the lower pool, where it is held for reabsorption as the soil begins to dry out. Materials to set up dozens of these simple gardens were relatively inexpensive and easy to procure. This keeps expenses low for the non-profit that grows produce for the local food pantry with the aid of volunteers. But the real beauty of this project is how replicable it is for individuals who want a small garden at home on their own roof or balcony. Many of the volunteers come with the desire to give back to their community, but also to learn from more experienced gardeners how to grow food in the city. The knowledge and simple techniques that they obtain at the Hell's Kitchen Farm Project can certainly help almost anyone to achieve a prosperous garden of their own. Find out more about the project at HKFP.org.

The third garden that I visited, located less than a block from Rosemary's, was the aeroponic garden on the roof of Bell, Book and Candle Restaurant. Aeroponic systems, also known as Tower Gardens, grow produce in a soil-free environment. Although the rooftop is not large, sixty tower gardens produce enough food vertically in three dimensions to supply 60% of the restaurant's fresh produce in season. Each tower is self-contained, automatically watered and fertilized, and easy to maintain without the risk of soil-borne diseases. The soil-free environment also creates a relatively light weight garden that grows an incredible amount of food without risking the integrity of the roof. Tower gardens have great potential for the home gardener, as well. The initial investment to buy a tower garden from TowerGarden.com is approximately $550 at this writing, plus the ongoing cost to purchase nutrients for the system. Although this is certainly more expensive than growing in a kiddie pool, the three-dimensional configuration of the tower garden provides much more growing area for roughly the same footprint. For the more frugal or handy gardener, a cursory internet search for DIY Tower Gardens will yield dozens if not hundreds of ideas to create your own vertical garden system. View some beautiful photos of the Bell, Book and Candle rooftop farm at bbandcnyc.com/photos/gallery/rooftop-garden.

These three outstanding gardens illustrate that one doesn't have to be a commercial farmer or a large-scale operation to grow food on the roof. Almost anyone can take advantage of the extra space on a rooftop, or even a sunny balcony, by growing food in a plastic pool, a small raised bed or a tower garden. And these are certainly not

the only options; the possible techniques are boundless, limited only by the creativity of the gardener. Read more and view pictures of the NYC rooftop gardens at the link in Figure QR 5.3.

INVENTING YOUR FARM

GROWING FOOD IN CONTAINERS

One popular and simple technique to grow food in small spaces is container gardening. Almost anything that you can grow in the ground can be grown in containers, with a little know-how. I personally love to grow in containers in what I refer to as my *patio farm*. Although I have plenty of space on my property to garden, my front porch and back patio are adorned by container plantings that add beauty to the landscape. Containers of herbs, peppers and onions grow right outside my door, serving as a kitchen garden that offers easy accessibility to items that I harvest most often.

To grow a patio farm, start by selecting the best containers for your plants. The following questions need to be considered:

1. What do you want to grow? In general, the pot ought to be slightly wider and slightly taller than the mature size of the plant you want to grow. Tiny succulents can grow well in coffee cups or shallow bowls, but a tomato plant will need a lot more space.
2. Drainage: The pot needs adequate holes in the bottom to allow for water to drain out of the pot.
3. Toxicity: Avoid growing edible plants in containers that are made of or once held toxic chemicals. Don't use treated wood, pallets or tires for edible gardens, as they contain preservatives and other chemicals that may be toxic to plants and harmful to humans.

As long as it meets these three criteria, anything that can hold soil can be a plant's home. Feel free to be creative and select containers that are both fun and functional.

POTTING MIXES: Once you have selected your containers, you will need to fill them with potting mix. A great potting mix is the foundation for a successful container garden. To make an effective potting mix, start with organic potting soil or compost.

To the potting soil, add some perlite and vermiculite. Perlite resists water and helps to maintain good drainage in your containers. Vermiculite holds water, and releases it back into the soil when conditions are dry. Mix one part potting soil, one part perlite and one part vermiculite to create your own all-purpose potting mix that is free of synthetic chemicals, that provides good drainage, and contains organic matter to provide your plants with nutrients.

Wet the potting mix ingredients and mix them well. Figure 5.4 reminds us that it is important to begin wetting the ingredients prior to mixing, as dust from the potting soil and perlite are not healthy to breathe. You will notice that it takes a lot of water to wet the potting mix thoroughly. The goal is to get the mix as wet as a wrung out sponge.

5.4

POTTING MIX

WET IT AND MIX THE INGREDIENTS WELL BEFORE PLANTING

BE CAREFUL NOT TO BREATHE PERLITE DUST WHILE YOU ARE POURING AND MIXING. IT CAN BE IRRITATING TO THE LUNGS

PLANTING: Gather your transplants and fill containers to within 1in of the top with potting mix. Make a hole in the soil for the plant that is as deep as the rootball of the plant, and slightly wider.

Gently turn the plant upside down and ease the rootball out of the nursery pot, being very careful not to damage the stem. Tease roots apart, snip off any circling roots and

place the plant in the hole. Be sure that the rootball makes very good contact with the potting mix, leaving no air gaps or pockets.

Return displaced dirt to the hole to fill in empty areas. Do not press or force the dirt into the hole. Gently water. The soil may settle, revealing low spots. If low spots appear, add a little bit more soil, as necessary, and water in again.

WATERING: In warm weather and dry seasons, containers tend to dry out quickly, particularly small pots and clay pots. Mulch the top of the soil with wood chips or pebbles to decrease evaporation. Water your plants until liquid runs out of the bottom of the pot to prevent mineral build-up in your soil. To facilitate draining, place the pot on small legs or in a shallow dish filled with gravel.

Allow the soil to nearly dry out before watering again. Roots need air, so a water-logged plant will die just as quickly as a dehydrated plant. Too much water also encourages the growth of mildews and molds.

FERTILIZING: Commercial potting soils generally have enough nutrients to last a couple of months. Following this initial period, add a balanced, water soluble organic fertilizer. For patio plants that are watered often, nutrients are flushed away rapidly. It may be helpful to refer to the directions on the fertilizer package, adjusting to fertilize twice as often at half the recommended strength. This provides a regular nutrient drip for plants that are contained and thus reliant upon the gardener to meet their nutritional needs.

ENJOY: Just about any container that can hold soil and water can become a plant pot, from buckets and boots, to reusable grocery bags and wine barrels. Have fun creating your unique patio garden, and enjoy the benefit of having your own fresh produce, right outside your door!

Detailed instructions and container gardening photos are located online at the link in Figure QR 5.4.

QR 5.4

PRODUCING FOOD OUT IN THE OPEN

From backyards and front lawns to patios and fire escapes, there is a growing worldwide trend to convert outdoor spaces that have been traditionally strictly ornamental to vegetable gardens and urban mini-farms. Often, these visible outdoor areas are homogenous, cookie-cutter spaces, where neatly trimmed grass or a few well-placed flower pots are admired and appreciated by the neighbors. But for some innovative gardeners, a feast for the eyes is not enough. They want something edible in return for the hard work, the water and the expense of tending a landscape. These food revolutionaries are maximizing their cultivation area by converting their landscapes, patios and nearby vacant lots into productive edible gardens. In the quest for more space to grow food, even conventional front lawns are being transformed into maverick, and highly visible, vegetable plots.

If you have considered converting your own front yard to a garden, how would it look? There is no single answer. Though publicity is given to highly energetic front yard farmers who convert entire lawns to gardens, not everyone is so ambitious. For some, it might be enough to grow a small herb bed or a border of edible flowers. Another approach could be growing a stand of fruit trees or a row of hanging baskets overflowing with strawberries. A few raised beds placed on the lawn would be another inexpensive and simple option. The configuration of your garden depends on how much time and money you want to spend and how you want it to look.

Another configuration concern is security. Front yard gardens are vulnerable in so many ways. You may need to take steps to protect your garden from issues that may arise. Perhaps you will need to build a low fence or plant ornamental bushes around the perimeter to deter pets and kids from accidentally damaging border plants. Low fencing that makes it more difficult to enter the garden, while leaving it visible and friendly to the neighborhood, may be all that's necessary to deter trouble.

On the other hand, you may wish to leave the garden completely open, inviting neighbors to peruse at their leisure and to partake in the harvest. If this is your preference, some damage or missing produce may be worth the price to foster a positive sense of community around your garden. Communicating your expectations with neighbors can go a long way towards heading off potential trouble. For instance, you may opt to tell the neighbors that they are free to pick produce, provided that they check in with you first so that they know what is currently ripe. Neighbors who understand your expectations, and who feel welcome in your garden, will likely keep a watchful eye on it when you are not around.

Speaking of the neighbors, how you design and keep your landscape will play a large role in whether or not they consider it an asset to the neighborhood. So, while a sprawling, disorganized garden may be just fine behind the house, a front yard garden has a greater cultural need for aesthetic appeal. Certainly, beauty is in the eye of the beholder, and the primary function of growing vegetables is to maximize the harvest. But if you desire to foster positive feelings amongst the neighbors concerning your garden, an attractive design is essential. Fortunately, an edible garden can be both gorgeous and productive.

Before planting, draw out your design and make a plan. Place tall and slow-growing plants behind the smaller, faster producing varieties. For instance, a westerly row of trellised tomatoes, interspersed with sunflowers, serves as a lovely backdrop for shorter squash and bush beans, and also provides some afternoon shade protection. Arrange the border with some small edible or ornamental flowers, and tuck in some herbs or onions to make full use of the space. To avoid bare spots and to extend the harvest, mix a few live transplants with some seeds of the same variety, so that when the transplants are harvested, seedlings will be growing up to take their place.

If visions of an edible estate are making you anxious to get started, be sure to check with your city codes and homeowners association (HOA) rules before tilling up the lawn. Some municipalities discourage, or even ban, front yard vegetable gardens. It would be a shame to go to the trouble of creating a garden, only to have to undo it later.

If this is the case in your area, you may need to get creative, perhaps growing a few unobtrusive vegetables in pots near the front door or berries in hanging baskets. Mix some herbs, greens and edible flowers discreetly into your garden beds. Replace a

shade tree with a fruit tree and grow rosemary or blueberries in place of shrubs. Plant sweet potatoes or oregano to serve as ground cover. There are many ways to expand your gardening space into the front yard without the telltale rows or raised beds.

QR 5.5

As the stories of renegade front yard gardeners are becoming public and interest in gardening grows, a groundswell of public pressure is rising in favor of gardening rights. Zoning ordinances across the country are gradually changing to support urban agriculture in its many forms. If your neighborhood doesn't allow front yard gardens, perhaps it may soon. In the meantime, grow food however you are able, and enjoy the benefits and satisfaction of having fresh produce from your very own yard. Read more at the link in Figure QR 5.5.

OH, GROW UP!

If horizontal space for gardening is lacking, grow in 3D. With a vertical garden, you can literally have food climbing the walls or windows.

One method of vertical gardening is to create a tower garden using repurposed bottles. Using a technique created by sustainable development expert Willem Van Cotthem, one can create an inexpensive garden in a sunny window, on a balcony or anywhere that space is limited. The technique essentially stacks empty food grade bottles, such as soda bottles or water bottles, attaching them vertically to a trellis. The bottles are filled with soil, and the system is watered via gravity from the top down. Easy to follow instructions are located at containergardening.wordpress.com/2011/09/07/bottle-tower-gardening-how-to-start-willem-van-cotthem.

Bottle gardening is just one method of growing food in three dimensions. There is a myriad of creative growing methods, and vertical gardening ideas abound online. Country Living online magazine's list of *19 Creative Ways to Plant a Vertical Garden* provides simple ideas to start growing vertically at countryliving.com/gardening/garden-ideas/how-to/g1274/how-to-plant-a-vertical-garden.

SPACE CONSIDERATIONS FOR LIVESTOCK AND POULTRY

RAISING FOOD IN 3D

Just like gardens, animals can be raised in 3D, too! Small livestock pens can be stacked in a manner that conserves space, saves labor and improves the health of the animals. At The Micro Farm Project, we raise rabbits in elevated hutches that are easy to move around. The rabbits spend time during our mild winters living in the garden above fallow patches. Their waste drops directly on to the soil, feeding the soil food web and preparing the ground for planting. Not having to scoop the waste and transport it from the hutches to the garden has saved me from being the middle man in the process. Figure 5.5 shows one of our rabbits enjoying some garden shade.

5.5

The rabbits were also briefly housed in our chicken run. The hens could run under the hutches, and they quickly broke down the waste into excellent compost. We moved the hutches out of the run because I was concerned about the chickens roosting on top and perhaps using them as a launch point to get over the 6ft fence. If our run were enclosed, this would be my preferred method of raising rabbits when the weather permitted.

AQUAPONICS

Aquaponics is aquaculture (raising fish) and hydroponics (growing plants without soil) in combination. This kind of small-scale, intensive food growing system can provide both vegetables and animal protein in a relatively small area. Waste from the fish is the perfect nourishment for plants, and it is pumped into a grow bed to the plant roots. The plants use up the waste and filter the water. Clean water is returned to the fish tank, and the cycle begins again. Plants grow very quickly in aquaponics systems. My friend, Chad Hudspeth of EndlessFoodSystems.com, gives an excellent description of how aquaponics works at youtube.com/watch?v=9idzIuo2m1g. The video is very informational, and although it includes marketing materials for Endless Food Systems, it is worth watching if aquaponics interests you. On a personal note, my daughters take care of Chad's goats, chickens, and his aquaponics systems while he is on vacation. I highly recommend that if you are setting up systems that include live animals that you find a reliable person beforehand who can care for them while you are away. It is so much easier to prepare in advance than to scramble at the last minute to find someone with both the skills and the willingness to help you out.

One interesting alternative to aquaponics uses duck pond water instead of fish tank water. Ducks can be very messy and spoil their pond water quickly with waste. Instead of dumping it out, the dirty water can be used to feed plants. Just as in aquaponics, the plants filter the water, and the clean water is returned to the duck pond. For more information, visit permaculture.co.uk/videos/how-create-aquaponics-system-ducks or permaculturenews.org/2012/08/22/quaquaponics-how-to-set-up-aquaponics-with-ducks.

TEENY-TINY LIVESTOCK

For a couple of years, I have dreamed of raising a cow. On my ¼ acre, I could probably get away with having one in the backyard, given the relatively permissive urban farming environment in Phoenix. The upside would be that I would know *exactly* where and how my steak was raised. The certain downside would be the enormous amounts of hay we would have to buy to support its nutritional needs. I am certain that it would be cheaper and easier to buy a side of beef from a local rancher. So, having a cow in the yard is definitely *out.*

But, wait! Just about the time I had given up on the idea, a solution was revealed by a book that I was using for research on miniature livestock breeds, *Microlivestock: Little-Known Small Animals with a Promising Economic Future.* The very first chapter was on (drum roll please …) *microcattle.* These are small cattle breeds that grow very quickly, have a high feed conversion ratio, can be milked, and are adaptable to arid climates, like ours. I did a little bit of searching and found that miniature cattle are available in Arizona, but at a premium price. A good friend recently purchased one and is raising it in a yard similar in dimensions to my own, so I will wait and see how it goes for her before I scrape up the cash to buy my own mini-cow.

Although having miniature cattle is still a *maybe* for us, miniature livestock breeds have been a key element on our farm. When selecting dairy goats, for example, we chose Nigerian Minis for their small stature and friendly nature. Easier to handle than full-size breeds, they seemed the right choice for our small children, who could have been potentially overrun by large Nubian or Boer goats. Had we been more forward-thinking, we would have selected miniature sheep breeds as well. We were fortunate that Oliver turned out to be a medium-sized ram whom I could control in an emergency. At one time, he escaped and threatened to ram our daughter Emily. I was able to side-step his charge and grab him by the horn to stop him, which I certainly couldn't have done had he been much larger. The same was true of our Mini Goat buck, Cassanova, whose friendly nature was sometimes overridden by his instinct to butt, which could have been dangerous if not for his diminutive stature. Figure 5.6 shows a picture of his handsome face and beautiful blue eyes.

Other than ease of handling, we discovered that Mini Goats have other benefits in a backyard farm environment. They require smaller living space than larger breeds. And although they produce a smaller supply of milk, they also require much less feed, so the cost to keep them is lower.

5.6

MINIATURE BREEDS

A GOOD OPTION FOR SMALL FARMS

SMALL BREEDS REQUIRE LESS SPACE AND LESS FOOD. THEY ARE EASIER TO HANDLE THAN FULL SIZE BREEDS

Similar to miniature cattle, Nigerian Mini Goats have a high feed conversion ratio. This means that the proportion of food intake is relatively small compared to the weight of meat or milk that is produced. This is extremely important on an urban farm where grazing is limited, and where some or most of the feed is purchased from exterior sources. On our farm, for instance, we have room to grow Moringa trees and fodder, and we can allow the goats to graze on weeds and landscape trimmings. Even so, we have to buy bales of hay, which are more expensive in Phoenix than in other parts of the country. Keeping expenses low is important to us, so miniature breeds make sense from an economic standpoint. Many urban farmers keep goats on properties that are smaller than ours. The smaller the space, the more important the feed conversion ratio becomes.

To learn more about the feed conversion ratio, visit The Pig Site at thepigsite.com/articles/4694/simple-calculations-feed-conversion-daily-gain-and-mortality.

Incidentally, this website is published by 5M Publishing, also the publisher of this book. The link will take you to an article concerning feed conversion ratios for pigs, but the concept applies to any sort of livestock. Miniature pigs are becoming quite popular amongst urban farmers, which is no surprise considering the popularity of bacon. One breed that I find intriguing is the Kunekune pig, which is adorable and known to be docile, quiet and easier to contain than other pig breeds.

Chickens, ducks and turkeys are smaller than their four-legged counterparts and often better suited to backyard farm environments. Keeping miniature poultry breeds may not be necessary under most conditions. But miniature breeds, known as *bantams*, can be particularly helpful in certain situations. If you have a lawn but are not in a situation to allow your birds to free range, three or four bantams could live quite happily in a mobile chicken coop, sometimes referred to as a *chicken tractor*. By moving the tractor around the lawn periodically, the chickens will enjoy fresh grass and will keep the lawn short. The adorable chicken tractor in Figure 5.7 is not mine, but the invention of Cricket Aldridge, blogger at GardenVariety.Life.

5.7

CHICKEN TRACTORS

MOBILE POULTRY HOUSING

UNOBTRUSIVE OPTION FOR SMALL FLOCKS AND SMALL FARMS WHERE FREE RANGING IS NOT POSSIBLE

For those who live in a community that prohibits livestock but allows pet birds, a visit to the state fair to buy registered show birds might fit the bill. These birds can be classified as pets. And if you get some of the more outlandish-looking bantam breeds, perhaps no one will even recognize them as chickens. Polish, Silky, Frizzle and Sultan chickens are fun to keep, entertaining and barely recognizable as chickens except that they lay eggs the size of a regular small chicken egg. Build the coop relatively small, below the fence line, and keep it clean. A friend of ours who lives in a planned community with a homeowner's association that prohibits chickens has been able to keep what he refers to as *tropical jungle fowl* in his yard, and many of his neighbors have followed suit. Because he keeps his property neat and his chicken coop clean, and has educated his neighbors on the benefits of keeping chickens, the homeowners' association (HOA) has now relaxed the rules concerning poultry. Check your local city codes and neighborhood rules to find out if you might creatively comply by raising rare breeds or registered show birds that could be classified as pets.

Other types of poultry also come in small or miniature breeds. Midget white turkeys are popular for backyard farmers, as are Muscovies, which are similar to ducks, but quieter and generally able to happily integrate with chickens.

If poultry of any kind are prohibited, even rare or registered breeds, another option for having fresh eggs is to raise Coturnix quail. Also known as Japanese quail, these birds are not poultry at all, and are generally classified as game birds. Quail must be raised in cages, and actually prefer small, enclosed spaces as long as they are not overcrowded. A rabbit hutch or aviary would be a suitable home for a handful of quail. They lay eggs very consistently when the days are long. The eggs are highly nutritious, one-third the size of a chicken egg, and people who have egg allergies can often eat them with no side effects. Even apartment dwellers who are allowed to have pet birds may be able to have a miniature egg farm on their patio or balcony by keeping Coturnix quail.

A PDF version of *Microlivestock: Little-Known Small Animals with a Promising Economic Future* can be downloaded at nap.edu/catalog/1831. The information is presented in a manner to assist economic development programs, but is useful for backyard farmers, as well.

CONCLUSION

We have discovered that, although finding space to produce food in the city can be challenging, it should not be an insurmountable obstacle. There are many ways to creatively grow and produce food, even in small spaces, if we expand our thinking beyond conventional methods.

Next, we will explore soil and its key role in food production. We will learn the key components of healthy soil and how urban farmers can transform bare dirt into thriving, productive humus.

FEATURED FARM
EDGE OF URBAN FARM

I visited with Scott Murray of Edge of Urban Farm, located in Vista, California. Scott and his wife Laura are producing enough food in a ¾-acre area to feed as many as eighteen families. What is quite interesting about the property is that it is situated on a north-facing slope, hardly considered ideal to a conventional agricultural model that prefers laser-level fields with a south-facing exposure.

I asked Scott how they were using the slope to grow so productively. He responded that when you are choosing an urban farm property, "… you often can't select the perfect place. Like this location faces north and it's a hillside facing north. So in the wintertime, we lose a lot of sun. Ideally we'd be facing south, where we had added sun in the wintertime because of the angle of the slope … And we felt we could make this work. And we found landlords that would allow us to develop it as a farm, and have animals … So we've adapted the property to suit our needs. We've done some terracing. We've done bed building. We've got pathways and walkways. We've made it as efficient as possible." Figure 5.8 shows a view of the top of the property from the middle of the slope.

I talked to Scott about how to start an urban farm and how to make really efficient use of space. "I got interested in urban farming because I wanted to farm and I couldn't afford a big farm. So the learning pathway is to make a small space very productive. And then if you have more space available you can develop a strategy, what I call a site specific strategy, that fits that space and allows you to optimize the production of that space and make a farm out of it." When asked how to come up with a site-specific strategy, he recommended starting with "… a blank piece of paper and a potential site that they know the boundaries of. You start by mapping, a simple perimeter map, and then you look at what features you have. And on an urban farm you'd be hoping that there was a water faucet on that property and that you had a water meter … a secure water source. Then you would assess the soil. Are you on a hillside, are you on flat ground, are you on rolling ground? How could we lay out a farm on this? And one of the most important things I

SEE THE POTENTIAL IN IMPERFECTION

PROPERTY DOES NOT HAVE TO BE IDEAL TO BE HIGHLY PRODUCTIVE

5.8

learned is, my first big farm acre, I was so eager to farm every square inch of it that I didn't leave big enough walkways at the ends of the beds."

When looking for an urban farm property, Scott recommends searching in certain areas of the city. "Now in modern development, they don't waste any space. They look at how many homes can we jam into this space, and the yards are much smaller. So I always recommend that people look in older neighborhoods and out on the edges. As the city grew early on, there (were) bigger lots. As the city grows now, development makes for very small spaces to farm."

I was particularly interested in a story that Scott recounted to me concerning an interesting way to farm, even if you don't own a single piece of property. "An example of site specific is a young woman that I met in Salt Lake City. She had gone four years of college, spent a whole year looking for work, could not find any work ... And she realized that she wanted to be a farmer. And

so she found (land) in her neighborhood in the avenues in Salt Lake. Her first year she farmed 2 acres that was made up of seven backyards. And in the first year she made enough money in the growing season to survive. And the next year she expanded to 4 acres and six employees. And now, seven years on, she farms almost 10 acres in the neighborhood in backyard patches, and ... does CSAs (community supported agriculture, farm ownership based on community members owning shares) and farmer's markets and sells to restaurants and to whole foods stores. And there is no farm. It's all little microchip-type patches. And her staff has these three-wheel cargo bikes ... so they ride from garden to garden, most of the time not even using a vehicle."

One last tip that Scott shared was to start a farm incrementally. "I think that the key for everybody is to start in steps ... And grow in increments because there are a lot of complicated pieces, moving pieces to a farm. And it's really good if your experience is pushing you up rather than you're stretching up past your experience and hoping for the best. And one of the key things to all farming is that the money you save is often the money you earn, that's your real profit."

Read the full interview and see more photos of Edge of Urban Farm at the link in Figure QR 5.6.

Chapter 6

Soil: *The Gardener's Most Valuable Resource, Right Under Our Feet*

It is wisely said that a garden will thrive when the caretaker ceases to nurse the plants and focuses on tending to the soil. A garden that is grown on the foundation of fertile soil will be more bounteous, have increased resistance to pests and disease, and, perhaps most importantly of all, will be stocked with abundant nutritional value.

And, *oh* – the smell of soil – like nothing else. One of my best childhood memories was the day we turned over the earth to start our garden in Kansas. I remember my dad dressed for gardening in his '70s-style short shorts, the hazy sunshine and how we sweated in the humidity. But most of all, I think about the smell and feel of the soil. I was not a kid who particularly liked to get dirty, but the moist earthiness of newly turned ground drew me to plunge my small hands deep into the cool mounds, turning over the soil to inhale the fresh, loamy scent.

Fast forward thirty-five years to the day that I started my garden in Phoenix. I wet the ground and ran the tiller. Disappointingly, no earthy smell wafted up from the earth. Hard, gray clumps of clay made the tiller bounce as clumps ricocheted off the tines with a ting-ting sound. It was obvious that this dirt resembled nothing like the loamy soil of the Midwest. Coughing in a cloud of dust, I muscled the tiller forward until it suddenly kicked back forcefully against what I thought was a piece of buried concrete. Later I learned that the hard, white material was actually a natural cement buildup of calcium carbonate, known as *caliche*. Upon further inspection, I realized that the concrete-like material was everywhere, impossible to dig out. But I am not one to worry much about details, so I made the decision to plant around it, for better or for worse.

You probably already suspect the outcome. The plants looked spectacular for a day or two before rapidly declining. I assumed that summer's dry heat was to blame and made desperate attempts to fix the problem with ever more watering. The season ended with a dead garden and a huge water bill.

I am almost ashamed to admit that I attempted a second garden using the same dirt as my foundation, this time in containers. Upon failing a second time, I did some research and discovered that in Phoenix, as in many places, healthy garden soil is not the natural state of things. Healthy soil must be fostered with diligence and copious applications of organic matter, which our clay soils lack. Knowing now what I didn't know then, I am fairly certain that my incessant watering had turned the pale dirt into something similar to modeling clay. Devoid of organic matter, the wet, compact earth literally smothered the roots of my plants.

I began to add compost to the garden, and covered the ground with a thick layer of straw mulch. Within a few short weeks, nearly all evidence of the compost and most of the straw had disappeared. I can only assume that the presence of water and food attracted starving microbes, earthworms and insects, who raided the new buffet with gusto.

It makes sense to me now that desert-adapted creatures would make quick use of any water and organic matter that becomes available since these are typically windfalls that arrive few and far between in the natural arid environment. Over and over, I amended the soil with organic matter.

Applications of water and organic materials were required much more often at first than they are currently. The reason for this may be, perhaps, that enough materials have built up in the soil to last longer. Or, it could be that a regular diet has made the soil organisms in my yard less voracious. Whatever the case may be, I now add compost twice annually, and lay down mulch once a year in late spring. My efforts are rewarded by ground that has gradually become dark, moist, and full of earthworms – all signs of healthy soil. The texture is crumbly, in contrast to the sticky clay that it once was. But the sweetest reward is in the robust growth of our plants that are supported by a foundation of healthy soil.

Building healthy soil does not happen overnight. But with patience and good practices, a gardener can create a living soil that holds the right amount of water, that cycles nutrients, that prevents disease and supports the growth of healthy, nutritious plants. This chapter is devoted to discussing a few of the many ways to build soil and keep it healthy.

In truth, the importance of building healthy soil cannot be overstated because it is directly responsible for the vitality of the plants that are growing in it. As we have experienced on our farm, it is an ongoing process, one that will be successful if we start with an understanding of the five components of healthy soil, and how to keep them active and in balance.

COMPONENTS OF HEALTHY SOIL

The first, and perhaps the most obvious ingredient is *dirt*, which is granulated rock that contains many of the micronutrients and minerals that are necessary to plants. These particles consist of sand, silt and clay. Relatively speaking, sand particles are large, silt particles are small and clay particles are tiny. If a single grain of each of these were placed side by side under a microscope, the sand particle's relative size would appear to be that of a beach ball, silt the size of a volleyball, and clay as small as a gumball.

Grains of sand hold together loosely, allowing for a great deal of air flow and water drainage in sandy soils. On the other hand, clay particles hold together tightly, slowing down the movement of water and hindering air penetration through the soil. To discover what kind of soil is in your garden, scoop a handful and get it wet. If you can form a snake or ribbon by rolling the material between your palms, the soil likely has a high clay content. If the ribbon falls apart quickly, you likely have sandy soil.

The second ingredient, *organic matter*, consists of fallen leaves, sticks, mulch and other decomposed materials that turn plain dirt into soil and give it a fresh, earthy smell. Some parts of the country have soil that is high in organic matter. Others, like the soil in Phoenix, contain almost no organic matter at all.

In order to sustain plants, soil also must contain the third and fourth ingredients, which are *air* and *water*. Air spaces are necessary for the health of plant roots and soil organisms. When soil becomes compacted, airways collapse and plant roots are smothered and stunted. Additionally, air spaces perform the dual function of transporting water. Water carries nutrients and supports life; not only plant life, but also other organisms that live in the soil.

The presence of these *living organisms*, both visible and microscopic, constitute the fifth component of healthy soil. Though mentioned last, they may be the most vital component of all. Without the presence of worms, bugs, bacteria and fungi, it is impossible for a garden to flourish. But when these organisms are in balance, nutrients are made available to plants, disease organisms are kept in check, and the plants themselves are tastier and more nutritious to eat.

SURPRISINGLY HARMFUL GARDENING PRACTICES

All five soil components must be in balance to create healthy soil. But two of the most common gardening practices that are intended to improve a garden are actually damaging to the soil. These practices are firstly, repeated tilling, and secondly, the use of chemicals, including herbicides, fungicides, fertilizers and pesticides. Though tillage and the application of industrial chemical concoctions promise a quick remedy to a troubled garden, repeated use dampers the activity of soil life that is necessary for the garden's long-term health. In effect, the productivity of the garden becomes dependent upon commercial products, labor and fossil fuels rather than the free energy of the sun and the soil food web. This is costly to the overall health of the plants, to the planet and to the gardener's budget.

When starting a brand new garden, tilling with a shovel or rototiller may be necessary in order to remove weeds, loosen soil and incorporate organic matter into the garden bed. But once the garden is established, repeated tilling can cause the ground to become quickly compacted, necessitating repeated tilling to keep the soil texture loose enough to allow air and water flow. This can become a never-ending cycle that disrupts the soil food web. Initially, when you add organic material to a garden bed and begin to water it, the soil comes alive with macro- and microorganisms that feed on the materials. These organisms begin to multiply rapidly, aerating and enriching the garden soil. Undisturbed soil teems with microorganisms and larger creatures, such as earthworms, which aerate and enrich the soil. Once your garden is established, tilling can disturb or even destroy this complex ecosystem, collapsing tunnels and damaging the structure of the soil. Additionally, tilling disrupts the immense networks created by bacteria and fungi that enable beneficial, symbiotic relationships with plant roots. Read my tips for soil building at the link in Figure QR 6.1.

Chemical fertilizers and pesticides are also detrimental to the soil food web. When a chemical fertilizer is repeatedly applied, plants will make use of the easy source of nutrition and lose their motivation to form mutually beneficial relationships with bacteria and fungi to meet their nutritional needs. Food web populations begin to dwindle. Lacking vital symbiotic relationships, the ability of plants to obtain moisture and nutrients from the soil declines, requiring increased applications of mineralized fertilizer and water to sustain them. The plants weaken, becoming intolerant of nutrient and water shortages, and the garden suffers unless both are continually supplied.

Additionally, when there is no longer any organic material for beneficial organisms to eat, they may become outnumbered by the pathogenic organisms that feed on the plants themselves. This problem is compounded by the use of pesticides that kill beneficial microbes and insects as well as the targeted pest species.

What is the solution to tilling and chemical amendments? Patience! Figure 6.1 shows that healthy soil is not created overnight. Compacted soil can be improved by dressing it with organic materials and watering deeply to attract soil organisms whose activities will begin to improve its texture. Prolific soil organisms also serve to make nutrients available, as well as to keep pests and pathogens in check, reducing or eliminating the need for chemical amendments. In Chapter 8 we will cover natural ways to treat problems in the garden – methods that do not harm the soil or create unintended problems down the road.

6.1

To learn more about soil, I highly recommend Dr Elaine Ingham's lecture *Life in the Soil, Part 1*, which you can view at https://www.youtube.com/watch?v=qXBIxFAxtlQ. You can also visit her website at SoilFoodWeb.com.

TOXIC SOIL

Growing food on commercial or abandoned lots is a wonderful way to find space for gardening in the city. One concern, however, is the potential for soil contaminants, such as lead, that could be present in the soil. The bad news is that human contact with toxic soil can present a health risk. The good news is that vegetables do not typically uptake these contaminants into their tissues, so the risk is easily reduced by limiting human exposure to the soil itself.

Placing a heavy layer of organic matter or mulch on top of the ground creates a physical barrier and humic acid present in the material binds the contaminants so that they are temporarily less mobile in the soil. Meanwhile, microorganisms activate to decompose heavy metals and chemicals. Additionally, a special kind of fungus, mycorrhizal fungi, begin to grow around the root systems of plants. These fungi trap and absorb heavy metals, preventing them from entering plant roots.

Building raised beds filled with uncontaminated compost and soil may be helpful for growing vegetables in a clean environment while the microbes go to work to detoxify the soil below. Additionally, cover paths with a heavy layer of mulch to reduce human contact with the soil and to encourage decomposition of toxins in walkways. It is recommended to wear gloves when working in the garden, and to wash vegetables thoroughly before eating. Root vegetables grown in soil that contains potential toxins should be washed and then peeled prior to eating.

QR 6.2

THE DIRT on farming

Fixing Toxic Soil

CITYFARMINGBOOK.COM/FIXING-TOXIC-SOIL/

Publications provided by the University of Minnesota and the Environmental Protection Agency provide a complete description of guidelines that will be helpful to urban farmers growing on potentially contaminated land. Learn more at the link in Figure QR 6.2.

COMPOSTING

Compost is decomposed organic materials that are tilled into the soil at the establishment of a garden, and regularly dressed to the top of the soil as a mulch. The addition of compost to the garden is a food source that attracts beneficial soil life. Compost breaks down quickly in the garden and must be reapplied several times each year, so many gardeners choose to make their own compost in order to avoid purchasing and transporting bagged products. It is very easy to create healthy compost, and there are many tutorials online that will show you how to do it.

The most common form of composting is *thermophilic*. This composting method breaks down organic waste using heat-loving (thermophilic) bacteria. One of the benefits of thermophilic composting is that it heats up when the microbes are active, and the resulting high temperatures kill diseases. The downside is that this type of composting requires a large pile, at least 3 square yards in size. If you have a lot of waste to compost, thermophilic may be the method to choose.

Compost is alive! Although it is made up of dead organic materials, living organisms that inhabit a compost pile do the job of breaking these materials down. Some of these organisms are visible, such as earthworms and ants. Others are invisible, in the form of bacteria and nematodes (microscopic worms). Others are easily overlooked, as in fungi that grow below the soil surface and sometimes are visible, in the form of mushrooms. These living organisms need nutrition, air, and water to thrive. So, it

is important to provide each of these three necessities in your compost pile. We will break down each step individually.

First, feed your compost pile. It needs a steady diet of "browns" and "greens". The simple recipe to follow is to feed the pile three parts "brown" materials to one part "green" materials.

This is not an exact process, so there is no need to measure the materials. Just remember to go heavier on the "browns" than the "greens". Here are some examples of "brown" and "green" materials:

Browns are organic materials that have been dead for a while. Often, dead materials turn brown, as is the case with fall leaves and pine needles. But they are not always brown in color. What they have in common is a high carbon content. "Brown" materials are most often dry, and may have been processed. They tend to decompose slowly, and are not typically attractive to pests, such as flies.

Greens are organic materials that were recently alive, as in scraps from a dinner salad. They are often fresh, as is the case with grass clippings. But, sometimes they are processed biologically, as is manure, or mechanically, as are coffee grounds. What they have in common is a high nitrogen content, which is a valuable food source for plants, as well as for the microorganisms that inhabit compost piles. They tend to have high moisture content and break down quickly.

Avoid adding meats, cheeses, and greasy foods to the pile as they may attract pests. Also, avoid dog, cat and human waste, although manure from strictly herbivorous animals is fine for composting.

Begin by building layers of brown and green materials. Start with a layer of *browns*. Layer brown and green materials, like building a lasagna. You can build the layers over a period of days or even weeks. Always top the pile with a layer of brown materials that will deter pests and prevent foul odors.

Once the pile has nearly filled the bin, begin to add moisture. Try to keep it as moist as a wrung out sponge, neither completely dry nor dripping wet. The addition of moisture will begin to activate the microorganisms in the pile, and their activity will cause the temperature to rise. The pile will become warm or hot to the touch. This means that your pile is breaking down quickly! Visit the pile every few days and put your hand into the materials to feel the temperature. When the pile begins to cool down, use a pitchfork or a shovel to stir or turn the materials so that some of the material that was at the center is moved to the edges, and the material at the edges moves to the center. Add enough water to keep the pile moist, and it will begin to heat up again. Repeat this process until the pile no longer responds to turning and watering by heating up.

When the pile no longer heats up, and the materials look and smell like soil, your compost is now *finished*. Stop turning and watering the compost pile and let it rest for a few weeks. This is called *curing* and it is an important step in the process. While the compost cures, it is breaking down any harmful bacteria that may be harbored in the material and converting nutrients into forms that plants are able to uptake and use. While your compost cures, you can start a second pile, if you so desire.

Now that you know how to create quality compost, let's put it to work! Start by sifting or picking out large pieces that have not decomposed sufficiently. Throw them back into your working compost pile to decompose some more. When starting a new garden, turn compost into the native soil. For an existing garden bed, work compost into the soil gently, using a garden fork. Compost can be added to potting mix for container gardens, and used to fill raised beds. Some of your cured compost can be added to a new compost pile to activate it.

Compost can also be used as a mulch layer that is spread on top of the soil like a blanket. The addition of a mulch layer does wonders for a garden, reducing weed growth and moderating soil temperature. It also helps to retain moisture by allowing rainwater to seep more easily into the ground and reducing evaporation. As the mulch breaks down, it becomes part of the soil profile, improving texture and nutrient availability. Add a layer of mulch seasonally, after planting.

Read more about composting on the Cornell Composting website at compost.css.cornell.edu/science.html. I also provide detailed composting instructions on the City Farming website at the link in Figure QR 6.3.

WORM COMPOSTING

What if you don't have space for thermophilic composting or winter temperatures slow decomposition to a halt? Worm composting, *aka* vermicomposting, can be accomplished year round in very small spaces, perhaps right under your kitchen sink! All that is needed to compost non-fatty kitchen waste is a small bin, shredded paper bedding, moisture and redworms. Provided with the right conditions, redworms will digest kitchen scraps and bedding very quickly, creating nutrient-dense manure, called *worm castings*. Castings can be harvested from the bin and used directly in the garden as a soil amendment. Explore worm composting at the link in Figure QR 6.4.

GARDEN BEDS

Perhaps we have gotten the cart before the horse, discussing compost before we have explored garden bed options. The garden bed is where you will put your compost to work. There are thousands of ways to grow a garden, limited only by the creativity of the gardener. Here are a few of the most popular ways to set up a garden bed.

RAISED BEDS

Raised beds are probably the easiest way to create a nearly instant garden. They require little soil preparation, just some weed pulling or debris removal to clear the area where the bed will be placed. The frame can be made from any number of materials, including untreated hardwood, cement block, bricks, or metal sheeting. Build the frame so that the sides of the bed are 8in to 2ft tall. The bed can be as long as you like, but no wider than you can reach across without walking in it. A width of 4 to 5ft is recommended so that you can care for the bed and harvest produce without needing to step on the soil. Line the bottom of the bed with carbon-rich materials, such as newspaper, cardboard, or straw. These materials will help to absorb water so that it does not run out of the bottom if the soil underneath is hard and water resistant. Over time, the carbon materials will break down. The native soil will begin to loosen up and incorporate with the soil in your raised bed. Fill the bed with quality compost, garden soil or potting soil. Now, you are ready to plant!

Raised beds made out of wood will last a long time if you select the hardest wood that you can afford and boards that are 2in thick. Thinner boards warp and crack very quickly. If you are not picky about the dimensions or the type of wood, hardware stores and lumber mills often sell scrap wood at a reduced price. If you will treat the wood for moisture resistance, be certain to use a non-toxic sealant. Figure 6.2 shows raised bed frames at The Micro Farm Project.

6.2

RAISED BEDS

BUILD A FRAME; FILL WITH COMPOST OR SOIL; PLANT!

USE NON-TOXIC MATERIALS, SUCH AS WOOD, BLOCK, ROCKS, METAL OR BRICKS

LASAGNA GARDENS

If space or energy are limited for gardening and composting, consider creating a lasagna garden. Also known as *sheet mulching* or *composting in place,* this method of gardening combines both in one location. It consists of creating a border for the garden and filling it with layers of compostable materials, rather than garden soil or finished compost. There is no need to till the soil beneath; organic materials are simply piled on top of the ground in layers of *browns* and *greens*. The materials break down as they would in a thermophilic compost pile, creating nutrient-rich compost in place with no need to transport them prior to planting.

If you are in a hurry to get your garden going and do not have the time or patience to wait for the materials to break down, build the layers of organic matter and cover them with 6 to 12in of cured compost. You can immediately plant a garden in the top layer of cured compost, which gives the plants enough soil in which to grow while the materials below decompose. I have created three separate gardens using this method. In each of them, the first growing season had fair success, with great improvement

THE DIRT

on farming

Sheet Mulching/ Lasagna Gardens

CITYFARMINGBOOK.COM/LASAGNA-GARDENING/

QR 6.5

in subsequent seasons. I have also noted that my lasagna garden beds hold moisture longer than beds that were filled with composted garden soil, which is a great advantage to me in our arid climate. Learn more about sheet mulching at the link in Figure QR 6.5.

KEYHOLE GARDENS

Keyhole gardens are all-in-one raised-bed planters and composting systems that look like a keyhole from above. They are built in a circular shape, generally measuring 6 to 8ft in diameter. The sides are constructed waist high with a cut away like the slice of a pie. The center of the circle contains a tower made of wire that serves as a composting receptacle. Keyhole gardens are particularly useful in dry climates because the soil mass surrounding the compost basket helps to keep the decomposing materials moist. In return, the compost releases nutrients and healthy microbes back into the soil. Besides using less water than conventional gardens, the height of the bed and the notch in the circle allow for easy accessibility to gardeners who would have difficulty bending down to an at-grade garden bed.

HERB SPIRALS

Herb spirals are an efficient, highly productive vertical garden design. The spiral shape allows gardeners to maximize space by growing plants in three dimensions. Spirals are typically built 5 to 6ft wide at the base, roughly 4ft tall with the center of the spiral at the top. The spiral surface provides a large planting area and varying microclimates to meet the water needs of various herbs. The spiral is watered from the top and moisture flows down to the bottom via gravity. Plants that require drainage and arid conditions are planted at the top, and those that need more water are planted at the bottom. The spiral design creates sunnier and shadier areas, as well as spots that are more protected from the wind. Thus, herb spirals are much more versatile than flat garden beds. Figure 6.3 shows an herb spiral at Roosevelt Growhouse in Phoenix.

HERB SPIRALS

COMPACT, SPACE-SAVING DESIGN

GROW FOOD VERTICALLY AND HORIZONTALLY TO MAXIMIZE SPACE

6.3

SOIL BUILDING WITH LIVESTOCK AND POULTRY

No matter what type of garden design you will employ, with the exception of hydroponic and aeroponic systems, healthy soil is necessary to support the plants. If you have livestock on your farm, you will naturally have plenty of raw materials with which to fill your beds.

Waste from livestock pens can be a source of pollution, not to mention odors and flies. Managed properly, however, animal waste can be a wonderful asset, providing materials for fast and effective soil building. Manure mixed with spent grains, spoiled hay, soiled straw and wood shavings can add up to the right combination of carbon to nitrogen that breaks down rapidly and results in healthy compost, rich in nutrients and microbes.

Waste from any vegetarian animal is suitable for composting. This includes goats, sheep, rabbits, horses, cattle and so forth. It also includes pets, such as gerbils, hamsters and guinea pigs. Poultry waste is also desirable, even though chickens are known on occasion to eat meat. Other than poultry, waste from carnivorous and omnivorous animals is typically not recommended, due to risk of contamination by E. coli and other pathogens. Some brave souls do venture to compost dog, cat and even human waste, both feces and urine. This is accomplished separately from other types of composting and is generally buried in deep pits. I do not have enough area on my property in which to bury it safely, and thus feel slightly *relieved* (pun intended) that it is not an option for me.

At The Micro Farm Project, three does and two sheep were more than able to create enough waste to provide compost for nineteen garden beds, with plenty left over to donate to the Mesa Community Garden. I once consulted with the City of Phoenix concerning management of the waste from our farm, and was informed that the city requires livestock waste to be bagged and discarded twice weekly. I understand why these rules are in place, but the idea of manure and straw fermenting in a plastic bag at the landfill seemed like a waste of excellent composting materials. A brief inquiry into composting rules yielded somewhat vague information, perhaps requiring a cement pad for livestock waste, and perhaps not. Since our composting system was working well, we quietly went on composting as we had previously, and the gardens improved year by year.

WORKING HENS

A bevy of hens can be a great asset to a gardener. Their scratching and foraging behaviors, and even their manure droppings, are a resource that no man-made implement or product can match. First of all, they are great at renovating soil and preparing garden beds. By allowing chickens into your garden between growing seasons, the birds will scratch and aerate the soil. They will eat all the dead and dying plant material, and will find pests and weed seeds that sometimes hide in the earth. Their manure adds nitrogen to the soil and increases microbial activity. Note that chickens should never be left unattended in any garden that is in production, as they will eat the produce, disturb the soil and potentially spread salmonella in their droppings. They do their best work in fallow beds between seasons or in new garden beds prior to planting.

Secondly, one of the easiest ways to compost is to employ chickens to do it for you. Materials compost quickly and thoroughly with the assistance of a few hens, which will aggressively scratch through the materials, searching for tasty morsels. This foraging behavior aerates the compost and breaks up the materials into small pieces. Their diet is diversified as they find scrap vegetation and insects in the compost pile, cutting down on feed costs and increasing the nutritional content of the eggs. Chickens digest these materials so quickly that they will often drop manure in the pile before they are finished browsing. The manure is high in nitrogen, which activates the compost pile to break down more quickly. It also serves to keep the pile moist. We throw all our compostable materials directly into the chicken run, minus avocado and the vines and leaves from any plant in the Solanaceae family, such as tomatoes and peppers, which are toxic to chickens. Composted materials build up quickly. Periodically, we dig compost out of the run, rotating the location in which we dig. Read about how we compost with chickens at the link in Figure QR 6.6.

Another method is to place a closed compost bin inside the chicken run, opening it when you want chickens to have access, and covering it when you want to keep them out. This is a much tidier method of composting with chickens, although materials that they kick out of the bin need to be raked and returned to the pile.

Dirty water from their drinking receptacles is a great resource for keeping a compost pile wet. It is wonderfully recyclable for composting purposes as it often contains bacteria and nitrogen that help compost piles to break down with lightning fast speed.

Thoughtful configuration of your coop, compost pile and gardens can facilitate the chicken–garden connection. One way to situate these elements is to place a south-facing coop between two garden beds, one on the east and the other on the west side of the coop. The compost pile can be placed behind the coop, between the gardens. Each season, one garden bed is in production while the other serves as the chicken run. So, if the east garden is in production, the west garden acts as a chicken run. The following season, the roles are reversed. In this way, crops are rotated, soil remains fertile, and pests and diseases are eliminated in a very simple manner. Compost materials are easily accessed by the chickens, and the location is convenient for transferring finished compost to the adjacent garden beds.

If space is limited, another option is to situate a bottomless chicken tractor and a few gardens in the same location. Build two or more raised beds whose walls are the same dimensions as your chicken tractor. Place the tractor so that it sits directly over one of the beds. The garden soil serves as the bottom of the enclosure. The chickens gain the benefit of living on soil without having to provide a large run for them. The soil, in turn, is aerated and fertilized by the hens. Rotate the chicken tractor amongst the beds seasonally.

GOAT ADVANTAGES

Ruminants, such as goats and sheep are fantastic for eating scrubby plants that are non-palatable to people or other animals. Goats are browsers, preferring to eat woody plants at eye level or above. They will chew on shrubs and branches happily. Sheep are grazers, more interested in grassy and leafy materials. Due to their extremely efficient digestive systems, ruminants are capable of converting nutrients from plants that people *don't* eat into delicious items that people *do* eat: namely milk and meat.

A wonderful by-product of their metabolism is nitrogen-rich manure and urine. Ruminant manure is a miracle soil amendment. It contains very few pathogens or undigested seeds, so it won't spread disease or weeds when you spread it in your garden. It is pelleted in small, compactly formed, relatively dry balls that release nitrogen slowly into the garden. Slow release of nutrients prevents that *burning* effect that high doses of nitrogen can cause in a garden. In my experience, it is advisable to bury the pellets in the soil, or at least under a thick layer of mulch. Pellets that are exposed to the air may dry out so much that decomposition stagnates, leaving hard, dry nuggets that remain on the surface of the ground, rather than releasing bacteria and nutrients down into the soil. For best results, keep them covered and keep them moist.

SPROUTED FODDER

If there is not enough space to grow your own animal feed in your gardens, go soil-less! Sprouting grain for livestock fodder is a highly efficient, easy way to add fresh feed to your animals' diet. It requires no soil or fertilizer; just a container with good drainage, water and seeds. Seeds are spread in a single layer in the container and kept moist until they germinate. After seven to ten days, the fodder is fed to animals whole; roots, leaves and any seeds that did not successfully sprout.

Sprouting is an extremely economical and nutritious way to feed livestock on a small farm. Germination increases the nutritional value of seeds, boosting the natural proteins, minerals, vitamins, enzymes, and omega 3 fatty acids. This amplification

of the nutritional value of the seeds results in a reduction in the amount of food that animals need to ingest in order to get the same amount of nutrition that they would from raw grains. Sprouting also has the outstanding ability to turn a pound of grain into 5 or 6lb of feed in just a little more than a week. Learn more about DIY sprouted fodder at motherearthnews.com/homesteading-and-livestock/sprouted-fodder.aspx. Figure 6.4 shows sprouted fodder trays at The Micro Farm Project.

6.4

SPROUTED FODDER

CAN INCREASE FEED WEIGHT UP TO 600%

SAVES MONEY, BOOSTS NUTRITION AND IMPROVES DIGESTIBILITY OF THE GRAIN

CONCLUSION

Although soil is not absolutely required to grow plants, as in sprouted fodder systems or aeroponic gardens, it is by far the most prevalent medium used for cultivation. In this chapter, we have discovered the five components of healthy soil and how to promote a thriving soil food web on our urban farms. In the city, although we may not have acres of ground stretching towards the horizon, we can make the most of our small spaces by fostering rich soil in garden beds and containers. The better the quality of the soil, the healthier our gardens will be, which will in turn support maximum yield and nutrition from our small intensive food production systems.

The following chapter will shift gears from fostering great soil to fostering new life on your farm. From seed saving and planting to animal husbandry, part of creating a sustainable urban farm involves perpetuating crops and flocks. Read on to discover how, following an initial investment, you can produce food year after year, without the expense of buying new plants or animals.

FEATURED FARM

SWEET LIFE GARDEN AND ORCHARD

I have had the pleasure of visiting some awesome urban farms and interviewing amazing farmers for the City Farming project. On June 15, 2015, Jill Green of Sweet Life Garden and Orchard gave me a tour of her property in Phoenix. This is not the first time that I have visited the garden, and I am always awed by its beauty and productivity. Even in the hot Arizona summertime, it is gorgeous (not at all crispy and brown, like my own personal garden.)

There is a lot going on at Sweet Life. With fruit trees and veggie gardens, goats and chickens, as well as newly added bees, Jill says, "It's almost like we don't need to go to the grocery store very often. And every time I go to the grocery store and look at the shelves ... it's all processed stuff. We have the fresh eggs, we have milk from the goats, and also we do meat birds one time per year. So we have our organic chickens. We have the bees, so we'll have the honey, too. It's a good feeling to be sort of self-sustaining, and then also to share with the community."

When I asked Jill her secret to having a lush garden in the desert, she recommended liberal additions of organic matter in the garden to hold precious moisture, and not just adding it to the surface of the soil. "I think for Phoenix, since it's so hot, the number one thing is to deep mulch. Don't let your soil get dried out; the deeper the better. We have our barn litter that we use in our garden. In our chicken pen, we put straw to keep the flies down ... And so, when we gather up the barn litter and clean up the chicken pen, we already have the nitrogen from the poop and the carbon from the straw. As we gather that up and bag it, we either compost it or put it straight into the garden beds ... The more you can layer on your leaves, or any kind of mulch, it keeps that soil nice and moist so that you don't have to water as much."

Jill has a specific method that she uses to layer organic materials in her garden boxes, which she learned after experimenting first with at-grade beds that did not thrive. "We read Lasagna Gardening, and I like the way that Patricia Lanza wrote that book. Before, all of our gardens had been in the

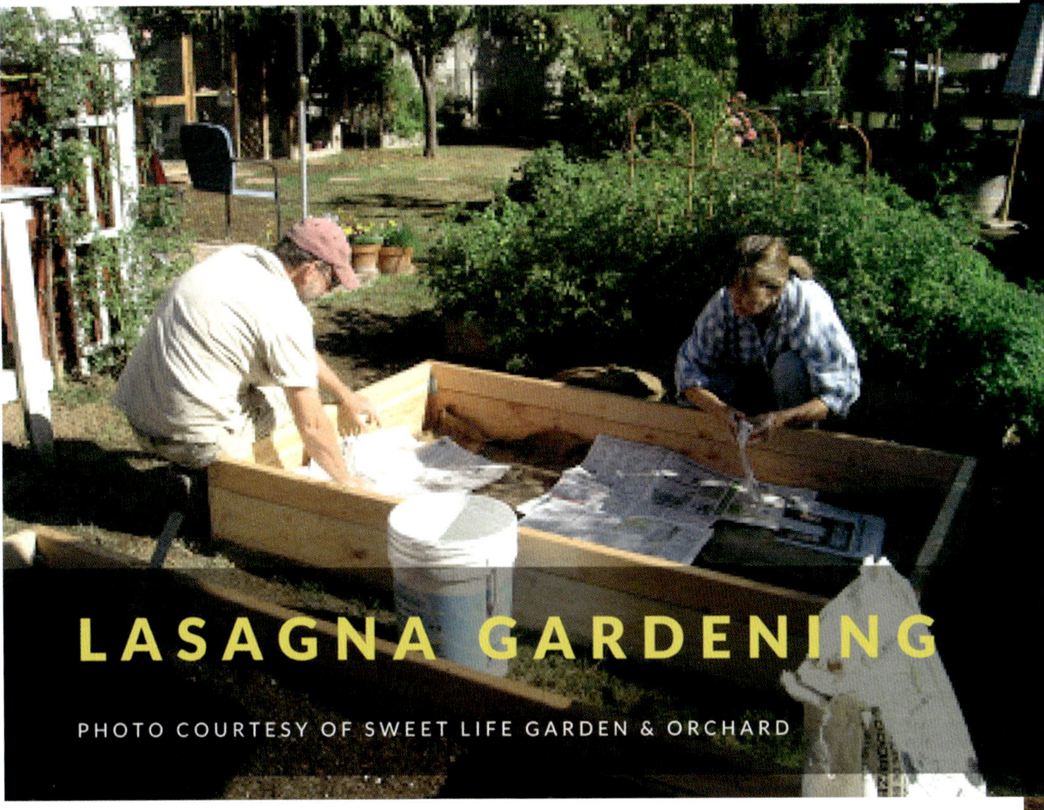

PHOTO COURTESY OF SWEET LIFE GARDEN & ORCHARD

6.5

ground and we would dump stuff in there (and dump and dump and dump,) and then we would plant. But it would just dry out, and there would be all these weeds and it was just not productive. When I read her book, I thought this was a good way to do it and we started doing the layering method. And then we also did raised beds (pictured in Figure 6.5) which, for us were so much easier because you can contain your soil and you can control it. And it gives almost as much produce as just laying it out in the (at-grade) beds."

Read the full interview with Jill and see photos of Sweet Life Garden at the link in Figure QR 6.7.

QR 6.7

Chapter

7

In With The New:

Propagating Plants and Breeding Livestock

Opening the barn doors wide, I breathed in the scent of fresh soil and clicked on the lights. The chickens rustled in the coop next door, and our curious dog came to inspect my unusually early activity. In the cool morning hour, I breathed the scent of damp soil and eyed my project. *Is it time yet?* I wondered. Over-eagerness could be disastrous, snatching defeat out of the jaws of victory.

A secret smile spread on my lips as my hand passed over the tops of the tomato seedlings, young but strong. This was the reward for so many weeks of work as scientist, farmer and mother of seeds. I had watched them grow, monitored moisture and temperature levels, carried them out into the sunlight during the warm days, returned them to shelter from the night frost. So far, so good … But tomato seedlings are finicky things, and I hoped the next transition would be successful.

Unseasonably balmy temperatures lured me to break ground and plant the seedlings in the sunniest area of our gardens. The plants would survive if the string of warm days continued. But a cold snap could spell their demise. I knew better than to trust the fickleness of Mother Nature, who could bring back the freezing temperatures just to spite my eagerness for spring. Even so, I had prepared the garden soil, and I was restless to plant.

The tomato cages were wrapped in clear plastic, ready to serve as support and frost protection for the young plants. I would have to trust them to do their job. Taking a deep breath, I loaded my garden cart with seedlings and supplies, and pushed out into the pink light of dawn.

A day of planting carried such promise, feeding the soul today and the body tomorrow. I dug a dozen planting trenches in the cool earth without gloves, gauging and enjoying the temperature and texture of the soil with my bare hands. Into each hole, I tossed a banana peel, a crushed eggshell, and a handful of alfalfa pellets. This ritual is my charm for a healthy tomato harvest, a bit of scientific magic that naturally meets the needs of my plants. As they slowly break down, the banana peel provides potassium and magnesium that help tomatoes withstand stress and resist disease. Eggshells release all-important calcium that prevents *blossom end rot*. And the alfalfa pellets contain nitrogen, minerals and triacontanol, a hormone that stimulates the growth of plant roots.

Performing another bit of magic to ensure a healthy root system, I pinched the leaves off the lower half of each plant and carefully laid them sideways in the trenches, burying the rootball and the leafless half of the stem under the soil. Though it pains me to entomb perfectly lovely stems with just the tops sticking out of the ground, the technique has the advantage of inducing tomato plants to grow extra roots from shoots that are buried in warm earth. Though the tiny, horizontally planted seedlings appeared as though listing sideways and drowning in a sea of soil, they would soon grow straight and tall. And the result would be bigger, healthier plants and a heavier harvest of ripe tomatoes down the road, thanks to a stronger root system. Tucking the plants in and labeling each unique variety, I set plastic-wrapped tomato cages upside-down to tent the seedlings. The effect was like rows of ghostly Christmas trees.

I stepped back to inspect my work, which was not too pretty, but effective for preventing damage in the event of a light frost. In a few short weeks, the plastic would be removed and the cages flipped over to support the growing tomato branches, which would surely be gaining both length and weight. I breathed a silent prayer for warm weather, brushed the dirt from hands, and began to consider what else I could plant …

WHAT'S RED AND GREEN AND STARTS IN DECEMBER?

Tomatoes in Phoenix, of course! The opening story illustrates some ways that gardeners overcome challenges to growing plants that require specific temperatures. I enjoy growing tomatoes, but the temperatures in our garden fluctuate between light frost and sweltering heat. Sandwiched in between the extremes are a few magical weeks in which the weather favors the ideal 70 to 80 degree range for tomato growth and production. Because tomatoes are slow growers and our tomato growing season is so short, seeds must be germinated indoors to give them a head start. The tiny seedlings require the most care and attention during the holiday season, between Thanksgiving and New Year's Day. Amidst all of the busyness of preparing gifts, processing turkeys, holiday decorating and harvesting bushels of citrus, I spend a lot of time coaxing tomato starts to grow in a sunny window. Starts made in late November and early December are transferred outdoors to the garden with frost protection in late January, or just prior to February 14. Yes, that is Valentine's Day in the United States, referred to as Tomato Day by gardeners in the desert Southwest, and it is regarded as the latest date to plant tomatoes if you want them to set fruit before it gets too hot.

Sometime in March, when the danger of frost has passed, frost protection is removed and we cross our fingers, hoping that the tomatoes are mature enough to flower during the spring, when temperatures are right for fruit set. As the heat begins to rise above 90°F, shade cloth is applied at a rate of 30% to create a delicate balance between decreasing the temperature just enough to extend pollination and allowing adequate sunlight for ripening. All this hard work is rewarded in May and June, when juicy, flavorful tomatoes are ready for harvest!

Thankfully, most garden plants are much easier to propagate than Phoenix tomatoes. Whether you grow from seed or transplants, planting a garden is a satisfying activity, full of expectation. This chapter will show you tried and true propagation methods that will ensure that your expectations are met with success. From the thrill of discovering seedlings that have germinated to gathering the harvest, it all starts on a solid foundation of proper propagation.

TRANSPLANTS

One of my favorite ways to spend a spring morning is to visit a nursery when the vegetable transplants have arrived. Transplants, also known as *starts*, are costlier per plant than seeds, but, for me, their value is not measured only in dollars. The instant gratification of selecting plants that are already partially grown and placing them in attractive garden arrangements makes the expense worthwhile. And for plants that require a long growing season, like cauliflower and celery, or that are difficult to germinate, such as thyme and flavored mints, I am willing to pay extra to have someone else start them in a greenhouse while I focus on propagating varieties that are easier to grow.

Successful transplanting starts at the nursery with careful plant selection. First and foremost, the varieties that you select should be in season. Although it may seem reasonable to trust the nursery to stock in-season items, it is not always the case. Therefore, the best tool to bring with you when shopping for starts is a local planting calendar, as discussed in Chapter 4.

Secondly, select plants that look healthy. That may go without saying, but I am a sucker for buying wimpy plants that have been relegated to the *clearance* rack, a habit that only occasionally turns out well. If leaves are brown, the stem is limp, or the soil has a mildew odor, move on to a healthier specimen. And if you can see roots growing out the bottom of the pot, the plant may be girdled or root bound, a condition that occurs when a transplant has been left too long in a small pot. Having no place to stretch out, burgeoning roots circle around the inside of the container. The result is a mass of matted root material that will never grow properly. It is nearly impossible to cure this condition, even by cutting out the circling growth. A proper transplant rootball contains enough roots to hold the soil together, but not so many that they are circling around the inside of the pot.

Once your plants are selected and brought home, it may be tempting to plant them out immediately. But wisdom dictates allowing the tender shoots some time to adjust to the conditions in your garden before subjecting them to the stress of planting. Help them to make the transition by gradually exposing them to the levels of sun, temperature, wind and water in your garden. This is called *hardening off*, and it gives plants that have been pampered at the nursery the chance to toughen up enough to withstand real garden conditions. Ideally, this is accomplished over a period of seven to ten days. Start by placing transplants in their nursery pots in the location where you will eventually plant them. If the change is very drastic from the nursery conditions to those in your garden, you might leave them there for only an hour or two, and then move them into a more protected environment. Gradually increase the length of exposure over a period of days. Once you are able to leave them in the garden most of the day without detecting signs of distress, they are ready to be planted.

Steps for successful transplanting:

1. Prepare a planting hole that is twice as wide and just as deep as the transplant rootball.
2. Use gravity to gently remove the plant from the nursery pot, protecting the stem from damage.
3. Place the plant in the hole, making sure that the bottom of the rootball makes good contact with the soil below. Backfill around the rootball with displaced soil.
4. Label the plant. I often forget this step, but it is very helpful to have a record of the plant variety, the planting date and the days to harvest.
5. Water well to soak the rootball and to settle the soil. Backfill with more soil, if low spots appear.
6. Cover the soil with mulch, leaving breathing room around the stem.

Figure 7.1 shows how easy it is to transplant starts. Voila! You now have an instant garden. Keep the soil moist for a few days, until the plants get more comfortable and established in their new home. You may need to temporarily shade them, too, if they have been positioned in direct sunlight. Shade protection could also serve as a buffer against driving rain or hail, which a tender seedling may not be able to withstand. Don't pamper them too long, however! Once they begin to show signs of growth, remove the shade protection and begin watering deeply when the top few inches of soil are nearly dry.

SEEDS

Though tiny, seeds are dynamos, loaded with energy and all the instructions required to create a brand new plant. In appearance simplistic and passive, these amazing marvels of nature actually have elaborate mechanisms that allow them to move, interpret environmental conditions, eliminate substandard specimens, detect light, sense gravity, defend themselves and pass on complicated genetic blueprints to the next generation. They have ensured the survival of thousands of plant varieties for eons, and are the custodians of the botanical (and culinary!) future.

Starting plants from seed empowers the gardener to take part in the miraculous perpetuation and curation of vegetables and other food varieties. A fascinating adventure, it is at the same time immensely economical and easy to do. The hardest part may be selecting which plants to grow. Seed catalogs often list dozens of varieties for any given vegetable, many of them beautiful or unusual. It can be difficult to decide which to choose! But whether you grow seeds directly in the garden or cultivate your own transplants, seeds can save money and increase production capacity. And, learning to grow from seed and to save your own seeds offers the satisfaction of self-sufficiently producing high quality food products from start to finish. Techniques may vary according to the type of plants you are propagating. Do some research to gather specific instructions for starting your particular varieties of seeds. These are the basics.

STARTING SEEDS FOR TRANSPLANT

Supplies:

- Seeds
- Seed starting medium (aka sterile soil)
- Planting calendar (for your region)
- Seed starting flats or trays
- Misting spray bottle
- Hand spade
- Grow light or sunny window

Instructions:

1. Test Seed Viability

Ideally, start with fresh seed. But if you have a packet of old seeds lying around, it may still be viable and able to germinate. You can test the viability of your seeds by folding a handful of them into a wet paper towel. Enclose the towel in a plastic baggie. Position the bag in a bright spot indoors for a couple of days and then unwrap the seeds to determine how many have sprouted. If most of them have sprouted, germination is good and the seed is viable. If only a few, or perhaps, none sprout, the seed may be too old or may have been stored improperly. Use a different batch of seeds, or sow heavily to make up for the poor germination rate.

2. Direct Sow

Seeds can be planted directly in the garden. Directly sowing seeds is simple and eliminates the need for seed starting trays, seed starting medium and eventual transplanting. The result of direct sowing is hardier seedlings that never experience shock or root damage that can occur during transplanting. Check your local planting calendar to know the best time to start your seed, and also research the types of seeds you want to grow to ascertain proper temperatures and seed planting depths.

As a general rule, seeds should be planted when the soil temperature begins to warm. Most seeds germinate best in temperature ranges between 65–80°F, but this is not always the case. Seed packets will often have temperature information printed on the envelope, and many charts are available online to help you to discover the ideal soil temperatures for germinating select varieties.

When planting, start with loose, level soil that is free of rocks and weeds. Seeds can be planted by poking them into the soil or placing them in a trench. Figure 7.2 shows seeds planted in a shallow trench. Plant seeds twice as deep as the diameter or length of the seed and cover with soil. Extremely tiny specimens, such a lettuce seeds, can be sown on top of the soil and covered lightly with a sifted potting mix or vermiculite. Water gently so as not to displace seeds, keeping the soil as moist as a wrung out sponge until they germinate. You may need to mist or lightly spray several times each day if the weather is dry in order to achieve the appropriate level of moisture. As seedlings begin to grow taller and develop a root system, you can begin to water more deeply and less often. In general, water as deep as the plant is tall to ensure that the entire root system is receiving adequate moisture.

7.2

To research proper soil temperatures for the seeds that you want to grow, read *Soil Temperature Conditions for Vegetable Seed Germination* at Extension.oregonstate.edu/deschutes/sites/default/files/Horticulture/documents/soiltemps.pdf.

If the soil is too cool for germination, cover it with black or clear plastic two weeks prior to sowing. Covering with plastic not only warms up the ground, but also forces pesky weed seeds to germinate. After the two-week pre-planting period, you will have a crop of tiny weed seedlings that can be hoed off to create a clean seedbed, ready for planting and free of weedy competition. Sow seeds and cover with plastic until the weather warms to the appropriate temperature for the varieties in your garden. If the plants begin to put on growth before you are ready to remove the plastic, cut an 'X' above each plant to allow it to poke through and grow unobstructed.

3. Grow Your Own Transplants

Some plants, such as peppers and tomatoes, need a long growing season to reach full maturity, longer than the average growing season in North America. Others have specific germination requirements that are difficult to maintain outdoors in the garden. These are often best started under controlled temperatures indoors. Begin with sterile seed trays and a planting medium that is free of insects, disease germs and weed seeds. Seed starting mixes can be purchased commercially, or you can mix your own. A good seed starting recipe is one-part sterilized compost or soil, one-part sand and one-part shredded sphagnum peat moss. If you prefer, substitute vermiculite or perlite for the sand. The idea is to create a seed starting mix that is light and porous to allow both oxygen and water penetration, which are essential for seedling survival.

It is also essential that the soil and containers are free of any fungal spores that could cause early termination of seedling growth. I have more than once been dismayed by seedlings that gave the appearance of strong growth, followed by sudden death due to fungus. This is called *damping off*, and avoiding its discouraging effects is the reason that using a sterile planting medium and clean containers cannot be over-emphasized. To sterilize your own soil or compost, moisten it slightly and place it in a cake pan to heat in the oven at 250°F. Using a candy or meat thermometer, heat the soil until it reaches and maintains a temperature of 180°F for half an hour. If the temperature starts to rise above 180°F, turn the oven temperature down. Avoid over-heating, as this can be extremely damaging to beneficial soil microbes.

Seeds should be started four to twelve weeks prior to transplanting, depending upon how quickly the seeds will germinate and grow. Avoid sowing seeds too early because holding them indoors too long may result in leggy, weak plants that are not hardy enough to survive transplantation.

To calculate when to start seeds indoors, check your local planting calendar to find out when it is recommended to plant the varieties that you want to grow. Mark the date on a calendar. Then, consult your seed packets to see how long it takes for the seeds to germinate and grow to a size that is big enough to transplant. Count backwards from the planting date that you marked on the calendar the number of weeks it will take to grow your transplants. This is your seed starting date.

4. Planting Seeds

To start your seeds, moisten your growing medium so that it is saturated but not dripping wet. Then fill your trays or containers to within ½ to ¼in of the top edge. Firm the soil with a wood block to create a flat surface, paying close attention to the corners and edges where soil tends to sink over time when not firmly packed.

Use an unsharpened pencil to create planting holes. Plant large seeds twice as deep as the diameter or length of the seed, two seeds per cell or planting hole. Extremely fine seed, such as lettuce and carrots, do not need to be buried. Lightly press them into the medium, cover with a thin layer of vermiculite and water in with a fine mist spray.

5. Water and Light

There are two ways to water seeds. Water from overhead by gently misting the seed containers, being careful to not flood or displace seeds. Or, water from below by placing the containers in a pan or tub that contains an inch or two of room temperature

water. Leave the containers immersed in the water until the soil soaks up enough moisture to be completely saturated. Then, remove the containers from the water to drain.

You will need to water often and monitor moisture levels carefully. Ideally, soil should remain as moist as a wrung out sponge so that it is wet enough to germinate the seeds, but not so wet as to deprive the seedlings of oxygen.

Seedlings need bright light to grow. Place them 4 to 6in below a grow lamp for about sixteen hours each day. Raise the lamp as the seedlings grow. If you do not have a grow lamp, situate them in a sunny window, if possible facing south. To increase reflection and the amount of light exposure that seedlings receive, cover a tri-fold presentation board with aluminum foil or Mylar and place it behind the seed trays. This will also serve to increase warmth in the window.

6. Thinning and Bumping Up

Seeds are planted two per cell or planting hole to ensure germination, but if both plants sprout, thinning is necessary. Once the seedlings have two leaves, select one seedling for removal to allow the hardiest to grow. Carefully cut it off at the base using a small snip or scissor.

If the plants have been seeded close together in one tray or in small cells, they must be "bumped up", or transplanted into a larger container to give them proper growing space. To know when to bump up, inspect the growth of the leaves. As seedlings sprout, the first two leaves are called cotyledons, or "seed" leaves. They may look different from true leaves. When the first true leaves appear above or between the cotyledon leaves, this is the proper time to bump up the plants.

To transplant seedlings in larger containers, carefully pry up the small plants, allowing them to fall apart from the soil and from each other if you have not thinned them. Carefully pick out individual plants, being mindful of treating the roots very gently as you disentangle them so as not to tear them. Poke a hole in the soil medium, deeply enough so the seedling can be placed at the same depth as it was growing in the seed tray. Holding the plants by the stem, plant them 1 to 3in apart. Lightly firm the soil and water gently.

Place them out of direct light and heat for a few days to help them get established, watering and fertilizing as needed. Some seed starting mixes contain a small amount of fertilizer. If your soil mix does not contain fertilizer, now is the time to begin feeding the plants with an organic, water-soluble plant fertilizer, at the dilution recommended by the manufacturer. Fertilize every two weeks until they are transplanted into the garden, being careful not to over-fertilize, which can be damaging to your seedlings. Learn more at the link in Figure QR 7.1.

QR 7.1

SEED SAVING: GROWING YOUR OWN SEEDS

For the gardener, seeds carry the power to economically create diverse and high-yielding food crops. They are easy and inexpensive to purchase. Nonetheless, astute gardeners do not rely solely on commercial seed, but carefully grow, collect and protect their own seed stores. Here are a few reasons to save seeds.

SAVING SEEDS SAVES MONEY

Though the price of a single seed packet may seem like an inconsequential expense, the cost to grow multiple varieties season after season adds up. By contrast, saving seeds is free. Growing and collecting seeds is simple and seed harvests can be copious, yielding plenty to save for your own garden with some left over to share or trade. I love to trade seeds with friends and at seed swaps; it's entertaining to discover varieties that others are growing and it adds incredible diversity to my garden – without spending a dime!

Seed saving also enables gardeners to cultivate stronger plants that are easier and more economical to grow. This is accomplished by selecting and planting seeds only from the healthiest, most productive specimens in the garden. By doing so year after year, the offspring will become stronger, more pest and disease resistant, and higher yielding. The need for chemical fertilizers and pesticides is reduced, thus decreasing the cost of growing food. For the natural gardener who does not use synthetic products, growing varieties that are pest and disease resistant is not only economical; it is crucial.

SAVING SEEDS CREATES WELL-ADAPTED VARIETIES

An additional benefit to selecting seed from the best plants in your garden is the creation of a customized seed collection that is well adapted to your growing conditions and to the specific cultivation methods that you employ. At The Micro Farm Project, water conservation is a priority. Only the most resilient tomato plants in our gardens can thrive during the hot summers on our frugal watering schedule. By collecting seeds from the strongest plants, we foster increasing drought resistance and heat tolerance in subsequent generations of tomatoes. We have a similar strategy for greens; those that are the last to become bitter and to bolt in the heat are the ones from which we save seed. By doing so, we are extending our growing season for kale and lettuces later into the spring.

And from the standpoint of personal taste, by selecting seed from plants with the best flavor, color and texture, you can begin to develop unique varieties that meet your own preferences. This is true for varieties that have been planted on purpose, as well as for *volunteers* that will sometimes appear in the garden uninvited. Early in our gardening adventures, a tomato plant grew in an area in which I had not planted, ostensibly from a seed that was hitchhiking in our compost. It turned out to be a very prolific plant that yielded loads of delicious cherry tomatoes. I saved the seeds and grew the plant deliberately for several seasons. I have to admit that I have stopped saving the seed as I have noticed that it self-seeds vigorously. I keep a few on hand, just in case the vine does not reappear of its own accord sometime in the future.

SAVING SEEDS PRESERVES GENETIC DIVERSITY

In the mid-to-late twentieth century, retail availability of seeds reduced the need for gardeners and farmers to save their own seed. Commercial seed producers naturally promoted their best sellers and proprietary hybrids. This led to a diminishing number of plant varieties that were commonly grown in home gardens and on commercial farms, and many heirloom varieties were lost or forgotten. Today, we see the effects of shrinking genetic variability, which is leading to decreased adaptability and reduced disease resistance in the plants on which we depend for food. Reduction of genetic diversity and adaptability amongst plants could eventually impair our ability to grow food. Thankfully, interest in heirloom varieties and seed saving is on the rise amongst both commercial farmers and home gardeners. By promoting and preserving diverse genetic variation, growers help to ensure the viability of future food sources.

INVENTING YOUR FARM

Here is how to grow your own seeds:

1. **Grow the right kinds of plants.** Three types of plants are available to home gardeners, but only two of three types will give you dependable results. In order to grow your own seed, you must grow the right kinds of plants. Here's the information on seed types.

   The first are *open-pollinated* plants. These are pollinated (without human assistance) by insects, wind, birds or other natural means. Unrestricted, random flow of pollen between individuals makes open-pollinated plants genetically diverse and adaptive. Seeds produced by open pollination grow *true*, meaning that the offspring plants will be nearly identical to the parent plants (as long as pollen is not shared between different varieties within the same species.)

   The second type of plant are *heirloom* varieties, which are open-pollinated varieties that have been passed down from generation to generation within a family or community. These plant varieties have been isolated and preserved for fifty years or more, and they often have an interesting story behind them. Heirloom varieties also grow *true*. One of my favorite heirloom varieties is Mrs Burns' Lemon Basil. As the story goes, organic gardener Janet Burns received seed for this variety in 1939 from her neighbor, who had been growing it since the 1920s. Notably, Janet's son, Dr Barney T. Burns, grew up to become one of the founders of a premier seed conservation organization, NativeSeeds/SEARCH.

   The third type is *hybrid* seed, which is pollinated by human manipulation under controlled circumstances, crossing two different plant varieties within the same species to encourage a desired trait. In the case of tomatoes, hybrids tend to be much more disease resistant than heirloom varieties. For this reason, I grow a mix of heirlooms and hybrid tomatoes to increase my chances of success. However, I do not always save seeds from the hybrid plants. Due to a phenomenon known as *hybrid vigor*, the first generation of hybridized offspring tends to be more vigorous and produce higher yields than the parent varieties. Unfortunately, the effect does not last. Seed produced by these offspring plants is genetically unstable and successive generations tend to deteriorate in vigor and productivity. Though it is technically possible to save and grow hybridized seed, the second generation will have non-uniform results, with a small portion of the offspring resembling the first generation plant, and many reverting back to the traits of the original parent plants. For this reason, gardeners who use hybrid plant varieties generally purchase new seed every year.

There is a fourth type of seed that is of growing concern amongst home gardeners: Genetically Modified Organisms, known as *GMOs*. Genetically modified seeds are produced in a lab by inserting genetic material from one species into another to produce a desired result. A well-known example is Bt corn, which is genetically altered to express proteins from the bacterium Bacillus thuringiensis that are toxic to corn worms. Crossing corn with an unrelated species of bacteria is very different from crossing varieties within the same species, as is the case in hybridization. Controversy has arisen regarding GMOs, and many home gardeners prefer to avoid GM seed. Fortunately, no seed companies currently market GM seeds to home gardeners. But if GMOs are of concern to you, informed vigilance is prudent as the situation could potentially change in the future.

2. **Grow healthy plants.** The key to producing high quality seed is to start by growing the most vigorous seed plants. Use your normal cultivation methods and refrain from favoring them with any extra water, fertilizer, shade or other special means. The objective is to grow seed that is well adapted to your *normal* climate and to the *typical* growing conditions in your garden.

As mentioned previously, some growers employ a different strategy, applying controlled stress in order to select the strongest plants. For example, in arid climates, exposing plants to controlled drought-like conditions and selecting seed from the plants that thrive under water stress will begin the process of creating a more drought-resistant variety for the garden. Or, if particular plants are commonly attacked by a pest, allowing the insects to invade the garden in order to select seed from the plants that resist the infestation creates a more pest-resistant variety the following season.

When employing stress to strengthen varieties, it can be tricky not to strain plants to the point that they will produce small, weak seed. When a plant begins to flower, scale back the stressors to which they have been exposed so that the plant is able to focus its energy on producing large, healthy seeds.

Collecting seed is very easy, but methods vary. Research your varieties concerning the signs that seeds are ready to be collected and seed harvesting techniques. Gather seed only from the strongest, highest yielding plants, and save only the largest, healthiest seeds. Discard small, underdeveloped or misshapen specimens. Allow the seeds to dry thoroughly, and store them in paper bags or glass jars. Keep the containers in an area that is cool, dark and dry, such as a cellar or refrigerator. This will ensure that the seeds will store well and will produce the most robust and vigorous seedlings possible.

If any of your seed plants show signs of disease, remove them from the garden before they begin to set flowers or they may cross-pollinate with your

healthier plants. If signs of disease appear after pollination, do not collect seed from sick specimens. Diseased parent plants can pass disease pathogens via seeds, infecting the next generation of seedlings.

3. **Make a plan.** Cross-pollination is the transfer of genetic information carried in pollen from one plant to another. This sharing of genetic material is a good thing; it increases diversity within a species and helps to ensure its survival. However, it may be undesirable if you are trying to save pure seed of a particular variety, such as a beloved heirloom, or if you intend to swap seeds with others.

   In order to avoid cross-pollinating heirloom varieties, it is important to understand which ones share pollen and which self-pollinate. Most plants cross-pollinate with others in their species. Pollen transfer is accomplished via wind, pollinator insects, animals, and so forth. Fortunately, some vegetable varieties will not cross-pollinate due to their nature of self-pollinating. Genetically pure seed is easy to grow with these plants. Self-pollinators include peas, as well as many varieties of beans, peppers, eggplants, potatoes, sunflowers and tomatoes.

   Other clues to understanding which plants will cross-pollinate and which will not are botanical names. These are the Latin (or Latin-ised) names of plants, which consist of two parts. The first name is a general (or genus) name and the second is a specific (or species) name. Together, the genus and species comprise the botanical name of the plant, written *Genus species*.

   Plants that share identical names may cross-pollinate, as in the case of muskmelons and cantaloupes that share the botanical name *Cucumis melo*. There are many members of the Cucumis family, but not all of them share the same last name *melo*. For example, the common pickling cucumber, although it shares the genus name *Cucumis* with melons, has the species name *sativus*, and thus will not cross with melons.

   Do not depend on common names, which can be confusing. For example, although pickling cucumbers do not cross with melons, Armenian cucumbers are of the species *Cucmis melo*, and thus will cross with cantaloupes and muskmelons.

   Map out your garden's layout to ensure that seed plants of the same genus and species are not planted too close together. Do some research to determine the recommended distance between plants or what barriers can be employed to prevent cross-pollination. To keep it simple, when I want to grow out plants for seed, I simply opt out of planting cross-pollinating plants in the same season.

4. **Water sufficiently.** At blossoming time, supply ample water that will allow plants to successfully set flowers and develop pollen. Insufficient water during this time period can impair seed yields and vigor.

QR 7.2

Once seeds have formed and are drying in preparation for dormancy, cut back the amount of water. At this stage, more arid conditions are preferable. Allow seeds to fully mature and begin to dry on the plant to enhance both their storage life and viability.

Keep a close eye on them at this stage, inspecting the plants often for signs that the seeds are fully mature so that you can harvest them promptly. Bring them under cover to finish drying, especially if rain is predicted. Repeated wetting of mature seeds on the plant delays dormancy and can damage seed tissues as they alternately shrink and swell. Mold and mildew may grow inside wet husks and pods, rendering the seeds unusable. Learn more about growing your own seeds at the link in Figure QR 7.2.

SEED SAVING: COLLECTING AND STORING SEEDS

When your garden plants are mature, the previous *INVENTING YOUR FARM* section recommended watching the flowers and the seed pods that will form in order to collect the seeds at the right time. How do you know when the time is right?

Here are some tips for collecting and storing seeds:

1. Explore harvesting methods. You will need to do some research on what you want to grow in order to learn specifics about when and how to harvest the seed. For most plants, harvesting techniques fall into two categories: wet and dry.

   Seeds that are harvested wet are found in juicy plants, such as tomatoes, eggplants and many melons and squashes. When the fruit reaches the peak of ripeness, harvest it and bring it inside. Allow the fruit to sit at room temperature for a few days, after which the seeds can be harvested. Swish the seeds in water and skim off the pulp and any seeds that float to the top. Viable seeds will sink to the bottom of the container. Strain off the water and spread the seeds in a single layer on a tray to dry. Pulp residue can cause seeds to stick to the tray as they dry, so a lining of wax paper or parchment is handy.

Seeds that are harvested dry are found in beans, okra, peppers, basil, as well as members of the onion and carrot families. Harvest dry seeds when the pods or husks have dried on the plant. If rain threatens as the pods are drying, shield them from moisture. Many seeds can be picked before they are fully dried and moved under cover, but you will need to research your specific varieties as some will not finish ripening after removal from the plant. In any case, it is preferable to leave seeds on the plant until they are fully mature and dry.

2. Place clean, dry seeds in a paper envelope, a Mylar bag or a glass spice jar and label them. It is important to include the date and the botanical name on the label, as well as the common name. Knowing the Latin name will assist you in planning your plantings for the next season. Check your seeds after a week or two. If any moisture has collected on the container, remove the seeds and spread them out to dry for a few more days before returning them to storage.
3. Store seeds in a cool, dry area out of direct sunlight. A dark pantry, a cellar or the refrigerator are good options. Place a silica gel packet inside the containers to absorb any residual moisture. A natural alternative to silica is to wrap a tablespoon of powdered milk in a double-layer napkin and place it in the seed container. Seeds that have been properly cleaned, dried and stored will stay viable for several years. Explore more information about seed collection at the link in Figure QR 7.3.

QR 7.3

THE DIRT on farming

Collecting & Storing Seeds

CITYFARMINGBOOK.COM/COLLECTING-AND-STORING-SEEDS/

STRATEGIES TO INCREASE AND EXTEND THE HARVEST

Grow more and grow longer in your garden by using four simple strategies. The first is to plant seeds and starts together at the same time. Set your transplants, and then immediately, or soon after, plant seeds around them. Your transplants have a head start on the seeds, and will reach maturity sooner. When the transplants have been harvested, the seedlings will be there to take their place. This is especially useful for lettuces, broccoli, cauliflower, cabbage and other vegetables for which you may want to harvest the entire plant all at once. Figure 7.3 shows transplanted cool season greens growing with seedlings sprouting in between the rows.

SUCCESSION PLANTING

SOW SMALL AMOUNTS OF SEED AT REGULAR INTERVALS

INSURES A PERPETUAL HARVEST OF FAST GROWING VEGETABLES

7.3

The second strategy is to combine large, slow-growing varieties with fast-growing smaller crops. The large crop will need to be spaced widely to accommodate its eventual size. Smaller crops can be grown in the empty spaces between each of the larger plants. At The Micro Farm Project, tomatoes are planted in January and February, while the weather is still cool. In between tomatoes starts, I sow the seeds of fast-growing greens, such as kale, lettuce and spinach. By the time the tomatoes are large enough to shade them, the greens appreciate the cover. Then, as temperatures begin to rise too high for the greens, I begin to sow basil seeds in their place. Tomatoes and basil naturally go together, and I love to have the staple ingredients for Caprese and Marinara sauce growing conveniently in one location.

The third strategy is to combine plants of different sizes and shapes. Tall plants and climbing vines grow well with low-growing varieties, and skinny shoots can be planted in between bushier crops. The effect is growing in 3D, which allows the gardener to maximize both the horizontal and the vertical space in the garden for greater yields per foot.

The fourth strategy is to use seeds to do same-crop succession planting. By planting starts or sowing small amounts of seed at regular intervals, fast-growing plants can be consistently harvested throughout the growing season. Interval plantings can either be made in the same bed or at scheduled times in different garden locations. For instance, planting a row or two of lettuce or radishes weekly ensures a continual supply for salads. For vegetables that do not keep well, this is preferable to a large surplus that is harvested all at once.

SEED LIBRARIES AND EXCHANGES

Libraries are great places to borrow books, movies, and *seeds*! That's right, seeds, and the libraries that lend them are popping up all over the country. How is it possible to borrow and then return seeds? The usual arrangement is to check out a packet of seeds, plant them, and let some of them grow to full maturity. The next generation of seeds is harvested and returned to the library for others to check out.

If you are interested in finding seed libraries, or starting one from scratch, Richmond Grows Seed Lending Library offers a wealth of free information and resources on its website: www.richmondgrowsseeds.org.

Similar to seed libraries, but a bit less formal, are seed exchange organizations. These groups get together to trade seeds, gardening supplies, and the fruits of their labor. Exchanges are diverse, with no single organizational blueprint. How each operates is up to the needs, desires and creativity of the members. Members get together in person or connect on social media. A great example is The *Veggie and Other Stuff Exchange* in Phoenix. The group started with a handful of gardeners who wanted to share their surplus. It quickly grew to more than 2,000 members on Facebook. Members are invited to organize and host their own exchange events, which are open to the entire group. Several exchange events occur each month, on varying days and in different locations, to make it easy for constituents to find an event that is convenient for them. Members trade all kinds of items, from seeds, tools and produce to jams, breads and fermented foods. A fringe benefit is a sense of community that develops amongst gardeners, and many lasting friendships have been formed amongst attendees.

STRATEGIES FOR LIVESTOCK AND POULTRY

Propagation strategies are not just for plants. Many of the same benefits that urban farmers get from saving and growing from seed apply to livestock, as well. If we think of offspring bred on site as analogous to growing from seed, and livestock purchased off location as analogous to growing from transplants, breeding on site is the clear economic winner.

1. Breeding on site is economical. Although having a rooster may be prohibitive on a city lot, males of many breeds of animals are legal, useful and desirable assets. It only takes one male turkey, quail, duck or rabbit to sire offspring from numerous females. In contrast to purchasing animals to increase your flocks one by one, the cost for on site breeding is free and numbers can multiply rapidly. Additionally, dairy animals need to be bred periodically in order to refresh their capability to produce milk. Having your own buck or ram is useful, if you have the space. However, even if you have to pay for off site breeding services, the cost is often worthwhile. Most pregnancies will result in multiple births, giving you the option to keep the females to increase the herd size or to sell all the offspring to recoup the cost of breeding.
2. Breeding on site perpetuates a herd or flock naturally. Animals come and go on a farm. Whether they die of natural causes or are processed for food, their life spans are relatively limited. Having the ability to replenish your stock organically eliminates the need to purchase stock and creates a greater degree of self-sufficiency on the farm. For instance, Cornish Cross chickens are the fast-growing meat bird of choice for many farmers, dressing to 4lb in as little as six weeks. However, they are hybridized for large thighs and breasts, so those that reach full maturity have difficult mobility and are unable to procreate. For this reason, some farmers prefer to raise meat birds that are able to lead more natural lives, such as Red Ranger broilers. Although these require a longer time period for growth, approximately twelve weeks to reach a 6lb dressing weight, they are able to forage, which reduces feed costs. What's more, they can reproduce naturally, significantly reducing the cost of replacement and dependence upon commercial breeders.
3. Breeding on site puts you in control of the timing. Spring is the primary season when young animals are available for purchase from feed stores and breeders. But spring is not the only time that animals may be needed or desired. In the case of Coturnix quail and meat rabbits that reproduce and mature quickly, these can

supply a year-round food source for the proactive farmer who breeds them on site. For farmers who supplement their income by selling stock, having animals available in off-seasons when competition is lower can also be profitable. Another consideration is reduced cost. At The Micro Farm Project, we used timed breeding to limit costs and management. In the case of hatching and brooding chicks, I found that if I hatched in the early spring when the weather was cool, the incubator had to work hard to maintain the required 100°F temperature. And once the chicks were hatched, a heat lamp was needed day and night to keep them warm enough. One fortunate year, a busy travel schedule prevented me from hatching spring chicks, and I subsequently discovered a better time of year for hatching in the Phoenix climate. Delaying until September, I set up my equipment in the garage. It is important to understand that the weather in Phoenix remains very hot into the early fall, and the high ambient temperature assisted in achieving the proper temperatures in the incubator. Once the chicks hatched, the heat lamp was only required at night after the first few days of life. This saved both electricity and effort on my part. Additionally, chicks brooded in the fall did not suffer through the stress of the summer heat early in life, but were able to acclimate to the outdoors and to living with the rest of the flock at a time of year when temperatures were mild.

4. Breeding on site puts you in control of promoting desired traits. As you gain experience as an urban livestock farmer, you will discover traits in your stock that you like and dislike, such as hardiness, disposition, udder size, feed conversion ratios, color, and so forth. By selectively breeding, you can promote favorable traits in the offspring, and discourage disadvantageous characteristics. Early in our farm experience, sentimentality caused us to make a mistake in this arena. We started out with two Nigerian Mini Goat does, Rose, who had long legs and Annie Oakley, who had short legs. Annie Oakley was the first of the two to kid, producing one doeling whom we named Calamity Jane. We fell in love with Jane and raised her to adulthood. Over time, I found that I preferred to milk Rose because her long legs made it easy to access her udder. Annie and Jane were more difficult to milk, requiring me to bend over in an awkward position. In hindsight, we should have sold Jane and kept Rose's offspring. But by the time that we discovered our mistake, sweet Jane had become a part of the family, so I put up with her *short*comings. Down the road, we had the good fortune to acquire Cassanova, a registered buck who came from a line of good stock, had blue eyes and exhibited a pleasant disposition. He sired many blue-eyed offspring that we registered and sold for a premium price and he earned his keep in providing breeding services for other farmers.

5. Breeding on site promotes hardy stock. In our experience with chickens, in particular, mixing breeds has resulted in stock with greater longevity and the ability to withstand the extreme Phoenix climate. We started out with purebred hens, with varying degrees of success. On a whim, I acquired some fertile eggs from our friends Dr Dave and Laura from Wish We Had Acres Farm. We successfully hatched a handful of them, and wound up with our own rooster, Kronus. The offspring of Kronus and the various breeds of hens produced what we lovingly referred to as our *mutt* chickens. Although these birds cannot claim any pedigree, they seem to be more robust and tend to outlive their purebred counterparts.
6. Breeding on site facilitates bonding. Gently handling young animals from birth fosters many advantages as they mature. Besides being friendlier, animals who have become used to human contact are easier to manage. For example, it is much easier to train a ram or a buck to walk on a lead when he is small than to attempt to do so once he has reached maturity. Our ram, Oliver, was held and brushed often as a lamb. As he matured, he never lost his desire for human contact, and would stand very still for brushing. This made it much easier to perform necessary grooming and medical tasks. He was also relatively easy to handle. He once escaped his enclosure and was preparing to charge at our daughter, Emily. I stepped between them, grabbed him by the horn, and led him briskly back to the pen before he had time to figure out that he could overpower me. Had we acquired Oliver as an adult ram, the situation could have been much more dangerous.

We discovered a similar advantage in poultry, which were easier to herd, to catch and to otherwise manage if we handled them often as chicks. Great examples of this concept were our roosters, Kronus I and his progeny, Kronus II. Our daughter Christina doted on them, carrying them around in her arms and on her shoulders as chicks. As they matured, they became aggressive towards strangers, and even towards my husband. But Christina never lost her ability to carry them around like babies, which she did often with both roosters and hens, as shown in Figure 7.4. In the evenings, the roosters had to be transferred from the coop to the shed in order to muffle their morning crows for the benefit of the neighbors. It was amazing to watch as the roosters would allow her to retrieve them from the roosting bars in the coop and carry them outside. From there, she

would set them on the ground where they would walk docilely to the shed, entering their kennels of their own accord. By contrast, Mondo was a bantam frizzle rooster whom we acquired when he was approximately two months of age and with whom we never successfully bonded. Though small in stature, he was much more skittish and difficult to handle than our larger roosters. This was perhaps due in part to inbreeding, but I believe that it had more to do with being kept in a cage and deprived of human contact from birth until he wound up on our farm. For this reason, though it may seem easier to buy a young animal rather than to raise one from birth, the bonding that occurs with babies is invaluable and may be difficult to achieve later in life.

7.4

BONDING

HANDLING ANIMALS WHEN THEY ARE YOUNG GETS THEM USED TO HUMAN CONTACT

THIS LEADS TO EASIER MORE PLEASANT INTERACTIONS FOR DAILY TASKS AND EMERGENCIES

HERITAGE BREEDS

Selecting livestock for an urban farm can be fun and exciting, but also confusing. There are numerous and interesting breeds of goats, sheep, poultry and pigs that fall into two main categories of livestock: commercial and heritage.

Commercial livestock breeds are the most common varieties used for retail purposes. These breeds have the advantage of very high production rates compared to the cost to raise the animal. Examples are cattle that gain weight quickly, chickens that lay eggs reliably, or turkeys that develop massive breast meat.

All of these benefits come with some serious drawbacks for the small urban farmer. Generations of breeding to enhance one particular trait often results in deterioration

of other desirable traits, such as declined disease resistance, diminished tolerance to heat and cold, curtailed natural life spans and the inability to reproduce or nurture their young naturally.

Lewis and I have experienced this first-hand on our farm. Our initial foray into raising turkeys started with Broad Breasted White and Broad Breasted Bronze breeds, also known as BBWs and BBBs. These animals are the same breeds produced by commercial turkey farms. They gain weight quickly, developing large breasts. Though the large amounts of breast meat are certainly desirable, the size of the breast makes it mechanically impossible for BBWs and BBBs to breed naturally. It became clear to us that raising these breeds would require purchasing new poults every year. Our interest was in maintaining our own flock, so the following spring, we purchased five heritage breed turkeys, Blue Palms. These beautiful turkeys were docile and friendly, and though they did not put on as much weight as their broad-breasted counterparts, they were able to reproduce naturally. In the fall, we processed three of the turkeys, reserving the largest hen and tom to parent the next generation.

As a result of our experience with turkey breeds, we became interested in raising heritage animals that were common in ages past before the rise of commercial breeds. We discovered that heritage varieties not only breed naturally, but also are capable of raising their young with little human assistance. A counterexample of this was our little flock of Coturnix quail, a variety of birds that have been bred for decades to enhance egg production, but which ignore their eggs. In order to perpetuate the flock, we had to incubate and raise the tiny quail chicks in a brooder. In comparison, our turkey hen was capable of brooding her own eggs, laying, hatching and raising a clutch of a dozen or more chicks at a time without any assistance.

Heritage breeds also tend to be very disease resistant, requiring few medications or vaccinations. They are frequently able to thrive in extreme climates, having originated in diverse regions of the world, from tropical climates to snowy mountains.

Generally smaller in stature than their commercial cousins, heritage breeds tend to be easier to handle and require minimal space for housing, which are great benefits to farmers in urban settings. They are thriftier eaters, able to put on weight on lower feed quality than their commercial counterparts can tolerate and are able to forage food for themselves. This is an important asset to urban farmers who must be creative in order to feed animals economically. On our farm, we raised Nigerian Mini Goats, which shared all of the above traits, able to withstand our high Phoenix temperatures

and tolerating all sorts of different types of feed, from fodder and Moringa to palm leaves and garden scraps. As a side benefit, these animals bred easily, and the money we made from selling the offspring was able to offset the cost of upkeep.

Groups of heritage breed enthusiasts exist for all sorts of varieties. A brief search on the internet will reveal information and advocates for all kinds of animals. To get started, check out The Livestock Conservancy at https://livestockconservancy.org.

HATCHING

If you are fortunate enough to live in an area in which roosters are allowed, you can perpetuate your own flock of birds by hatching fertile eggs. Often, a broody chicken will accommodate, sitting her own eggs as well as those of less broody hens. Turkey hens will also hatch eggs. In our experience, it is wise to separate broody turkey hens from toms who may kill the hatchlings.

Hens often collect a group of eggs, known as a *clutch*, prior to beginning to sit on them. This can take a period of days or weeks. Once she begins to sit, she will stay in place except to eat and drink. Her body warms the eggs, signaling the embryos inside to begin developing. Chicken eggs take twenty-one days to hatch. Turkey and duck eggs require a longer incubation period, about twenty-eight days from the time that the mother begins to sit to the hatching date.

If you do not have a broody hen, or if you are raising Coturnix quail, who do not tend to sit their own eggs, you will need an incubator to provide the right environment for hatching. Incubators range in price from expensive commercial apparatus to cheap systems for home users. I recommend buying a home system with a forced-air fan to circulate warmth throughout the incubator. In my experience, these work much better than those that employ a heat source without a fan. In order for embryos to develop properly, eggs must be turned daily. An automatic egg turner makes this job much easier and is worth the expense, in my opinion.

Find out how to hatch using an incubator by following the link in Figure QR 7.4. Once the eggs have hatched, chicks can stay in the incubator for a day before moving them into a brooder box. The box keeps them safe and warm under a heat lamp

while they grow and develop true feathers. It is easy to set up your own brooder box. A plastic tub or cardboard box will do. Line the brooder with shredded paper or a beach towel. Provide a water source that allows the birds to drink, but does not allow them to step or stand in the water. Chicks require a starter feed that is made specifically for young poultry. Cover the box with a screen that protects the chicks while allowing plenty of airflow.

To keep the tiny birds warm, suspend a heat lamp over the brooder box. They will need a heat source until they are about five weeks old, at which time they will be fully feathered and able to regulate their own body warmth. If chicks appear to be cold, huddling under the lamp, move it a little bit closer to the box to increase the temperature. If they appear to be hot, avoiding the lamp, raise it slightly up and away from the box to decrease the temperature. More information about brooding chicks is posted at the link in `Figure QR 7.5`.

Incubated chicks can be introduced to the adult flock when they are roughly the same size as the mature birds. Be prepared for some turmoil in your flock while the younger and older birds get used to each other. Aggressive birds may squabble and pick on the new birds until they establish a new pecking order. You can minimize the agitation by introducing the birds strategically. When chicks are large enough to move out of the brooder, but not large enough to join the flock, provide an area in the coop or chicken run for them that is sequestered so that the birds can see, but not touch, each other. In this way, they will get used to each other's presence without the danger of bigger birds harming the smaller ones. When the chicks have grown to a similar size as the mature hens, remove them from the sequestered area and place them with the adults on the roosting bars at night when the birds are going to sleep. In the morning, give everyone some extra treats or allow them to free range in order to distract

them as much as possible from squabbling. There will likely still be some aggressive behavior, but these steps will help to facilitate a smoother transition. Explore more about keeping chickens at the link in Figure QR 7.6.

CONCLUSION

Propagation skills require time, practice and education to develop. But it is well worth the effort in terms of cost savings and the increased hardiness of both plants and animals. It is also highly satisfying to produce one's own transplants and succession livestock. In truth, a farm cannot be considered truly sustainable or regenerative unless it is self-perpetuating.

As you organically grow your farm, setbacks may occur in the form of pests and disease. In the next chapter, we will explore natural methods to protect your farm, as well as ways to manage problems as they arise. We will also discover how to produce food that is nutritious and free of harmful chemicals in a healthy, thriving environment.

FEATURED FARM

THE YOUTH FARM

On a cool April day, one month prior to the peak growing season, Lewis and I visited one of NYC's most visionary urban farms, The Youth Farm located in Brooklyn. The farm was started in 2010 when the founding Principal of the High School for Public Service in Crown Heights conceived of converting the fifty-year-old lawn into an outdoor classroom. To accomplish the task, he partnered with BK Farmyards, a local urban farming organization, to develop farm plans. According to The Youth Farm website, "Ultimately, the school voted to have the lawn converted into a productive farm, staffed by trained farmers, that would provide the school and surrounding community with opportunities to engage in growing their own food and cooking with it. Since 2011, the farm has thrived as one of NYC's largest urban farm sites, providing training opportunities to youth and adults, increasing options for fresh, affordable produce, and creating a vibrant community space that cultivates radical change from the inside out."

The school currently provides youth education, as well as leadership programs, farm internships for young adults, community volunteer days, tours and workshops. It produces food and flowers for a farmer's market, CSA and farm-to-restaurant program. During our visit, adult interns were at work, preparing for the imminent growing season. The farm utilizes a number of strategies to extend the growing season, which is relatively short due to persistent cold temperatures in the coastal regions of the northeastern United States. Simple box cold frames bordered garden rows and a hoop house stood at one end of the property.

Production is boosted by combining propagation strategies. Several long rows of cool season vegetables had been direct sown in at-grade garden beds. Additionally, interns were carefully seeding trays to grow transplants for the farm, shown in Figure 7.5. Precision was aided by a detailed log of what was being planted, as well as the use of pelleted seeds. Pelleting is a treatment that coats small or irregularly shaped seeds with an inert material to make them large and uniform. The increased size and weight of pelleted seeds makes them easier to separate and plant. Deliberate placement of seeds

7.5

with precise spacing renders thinning unnecessary and allows roots to grow unimpeded by competing plants. Additionally, some pelleted seeds are primed to increase and speed germination rates. The co-manager of the farm, Molly, explained that pelleted seeds are more expensive than typical seeds. However, the precision and fast germination that they afford can save a farmer or gardener money in the long run.

QR 7.7

Hundreds of healthy plant starts in varying stages of growth soaked up the sun on tables in the farmyard, and still more were enjoying the warmth of the hoop house. Although I would have loved to have seen the farm during full summer production, the sight of the vast number of green starts offered the promise of a productive future for the farm and the hope of a brighter future for the young farmers.

Read more about the Youth Farm at the link in Figure QR 7.7.

Chapter

8

Two Steps Back:

How to Overcome Setbacks from Pests, Diseases and Toxic Chemicals

In the month of March 2013, I had grown enough Red Russian kale for my family to consume every single day throughout the winter. And I had learned something – *Everything gardens.*

For months, we had eaten kale salads of all sorts – sautéed kale, kale smoothies, soups and sauces made with kale, kale crostini, kale chips … and we were getting a bit tired of it. The end of the kale growing season was approaching, and I was relieved. In truth, I had ignored the kale patch for several days, turning my attention to planting summer vegetables.

Even so, I wanted to harvest enough of the last bits of kale to dehydrate for our summer smoothies. To my chagrin, I discovered that the crop appeared to be wilted and covered with a fuzzy, sticky substance. Upon closer inspection, the substance was revealed to be a common garden pest known as plant lice or aphids. Aphids are nasty little soft-bodied insects of several varieties, each about the size of a pin head. One aphid is virtually unnoticeable. Unfortunately, these tiny sap-suckers make their appearance in hordes that number in the hundreds or thousands, and their presence rapidly drains the life out of garden plants via group effort.

My second unpleasant discovery was a proliferation of ants; in whose path I was apparently standing. Jumping away from the anthill, I quickly grabbed the garden hose and turned the nozzle to a hard spray, angrily delivering a blast of water to the ants that were biting my legs and subsequently to the aphids inhabiting my kale. Laying the hose down, I leaned over and inspected the dripping plants. Most had been rendered inedible; a few were relatively unscathed, showing only the first signs of infestation.

And so the war began to salvage what was left of the kale – Kari armed with a garden hose and soapy spray; the aphids armed with tenacity and overwhelming numbers. What I did not consider when I undertook the battle was the force of nature, which was clearly on the side of the aphids. As it turns out, aphids play a key role in nature's clean-up crew, and are evidently commissioned to speed the decomposition of weak, sick or out-of-season plants. The infested kale didn't stand a chance, despite repeated blasts with my soapy spray. I made a mental note to avoid the wasted effort in the future.

Singularly focused on my battle with the aphids, I almost missed the significance of the ants. Initially chalking up the simultaneous aphid–ant infestation to an unlucky coincidence, my curiosity was piqued when I noticed that the ants seemed to be drawn to the aphid-ridden area of the garden. I wondered if there was a connection. A cursory internet search revealed the fascinating truth that aphids essentially poop sugar (aka *honeydew*). And what do ants like best to eat? Sugar!

Figure 8.1 shows ants and aphids feeding together. Even more fascinating is the fact that ants farm the aphids, corralling them like mini-ranchers and guarding their herds from predators. Amazingly, dairying ants don't just feed on the honeydew left behind by the aphids, but they actually *milk* the aphids by stroking them with their antennae. Imagine that.

8.1

APHIDS

FEED IN CLUSTERS ON NEW PLANT TISSUES, CAUSING STUNTED GROWTH OR DEATH

HONEYDEW EXCRETED BY APHIDS ATTRACTS ANTS & ENCOURAGES GROWTH OF BLACK SOOTY MOLD

The ants' similarities to human farmers don't end there. Just as farmers store up seeds, some ant species harbor aphid eggs in their nests over the winter and protect them while they hatch. In the spring, they deliver the hatchlings back to the plants or to plant roots that grow underground within the ant colony. And when a queen ant leaves to start a new colony, she may take an aphid egg with her to establish her own herd. Seriously!

The ant–aphid connection drove home to me the meaning of one of the Attitudinal Principles of Permaculture – *Everything Gardens*. In other words, every creature engineers the environment to create conditions in which it can thrive. And in the interconnected web of life, each creature carves out its own niches. At first glance, the situation in my garden seemed simple – just a bunch of aphids attacking my kale, and ants being an annoyance. But further investigation revealed an extremely intricate, complex system that was at work in my garden's mini ecosystem, and uncovered some fascinating information about the aphid–ant connection. But, more importantly, it revealed how little I knew about the workings of nature – and how ridiculous it seemed to combat forces unknown and unseen with a bottle of soapy spray.

Those of us who were old enough to watch television in the 1970s may remember the iconic Chiffon Margarine commercials. The ads starred Mother Nature, who tasted the margarine and believed it to be butter. When she was corrected by an anonymous narrator, her memorable tagline was, "It's not nice to fool Mother Nature," followed by ominous thunder and a flash of lightning. In one version of the advertisement, Mother Nature glided on to the screen on a vine, like Tarzan. The unseen narrator remarked, "I didn't know you were so fine on the vine," to which she replied offhandedly, "I bet there's a lot you don't know."

The reality is that you can't fool Mother Nature. She may be circumvented for a while, but nature always wins in the end. And the profound truth is that there really is *a lot we don't know*. The more that science reveals, the more we discover that there is yet to discover.

So, what's a gardener like me to do? The knowledge that I may never fully understand all the intricate workings in my garden sounds like a serious setback, a disadvantage that might be impossible to overcome. But I see it as an asset. Humility makes me a *better* gardener. It slows me down, makes me think and investigate, to consider the entire system that is at work in my garden before choosing a course of action. And it gives me a really good reason to work hand-in-hand with natural processes to the best of my ability, instead of fighting against them.

I have difficulty with considering the entire system as a whole, aka *whole-systems* thinking. It overwhelms me. Perhaps I am thinking *too* hard about it. Fortunately, people who have been practicing Permaculture and whole-systems design (much longer than I) came up with some tools to simplify the process, the Permaculture Design Principles. These principles are thinking tools, that when used together, help us to observe what is happening in natural ecosystems and to mimic it. By mimicking what nature does, we can create systems that run more easily, with fewer problems and mistakes. They are not only good for the environment, but also make life easier. Here are a few that I have found to be particularly helpful.

PERMACULTURE PRINCIPLES

OBSERVE AND INTERACT

The first principle is *Observation.* This principle can be applied to all aspects of life, not just gardening. It is said that Permaculture is 90% thought and 10% work, i.e. brief periods of activity that follow protracted observation. This runs counter to our culture, which often demands a quick response and tends to spend large amounts of time in activity and little in contemplation or observation.

Taking time to ponder and observe helps to ensure that when we do finally act, our actions will be appropriate and effective. This is true for almost every aspect of our lives, not just gardening. In the area of finances, money is handled more efficiently when income and outgo are budgeted and tracked to see where improvements can be made. And in business, systems are studied in order to improve their efficiency.

Observation is a particularly beneficial exercise for urban farmers. It includes passively watching what is happening on the farm property, as well as active investigation and learning. Both are important. Passive observation occurs by regularly spending time on your property and focusing in on what is occurring there. It may include noting weather and water patterns, wildlife, plant growth, human activity, and so forth.

When something interesting is noted, passive observation may become active. I have a weed that pops up in my yard that is called Puncturevine. The seeds make a large thorn, often referred to as a "goat head". The thorn is very sharp, dangerous to animals' feet. It will poke through the bottom of a sandal and even puncture bicycle tires. For many years, I hated to see it in my yard. Passive observation taught me that this particular weed would rear its ugly head in the springtime. My stress levels would rise when the weather warmed and the weeds started to appear, and I was quick to eradicate the Puncturevine as soon as it sprouted. But my attitude towards it changed when I began to do some active investigation, attending an herbal tincture class to learn about making natural, homemade remedies. The instructor began to educate me on the dietary and medicinal uses for weeds, and it dawned on me that the weeds in my yard could have value other than as fodder for my chickens. It also occurred to me that weeds grow in my yard without any assistance. What could be better to a gardener than a crop that requires no planting, watering, or pruning? The tincture class, coupled with my own observational experience, began to change my view of weeds in general. And what plant do you think that we used to make our first tincture? You guessed it ... Puncturevine! Figure 8.2 shows my daughter and me showing off the tinctures that we made in class.

Emily & Kari at Tincture Class

8.2

I never viewed that weed in the same way again, and the next time I saw it in my yard, I was happy to collect and use it as a medicine. In the years following the tincture class, other common weeds have become prized plants on our property. I have learned to sauté purslane for its high Omega 3 fatty acid content and lamb's quarters for beneficial minerals. The Micro Farm Project Garden Club harvested dandelion both for salads and for tinctures. Taking the time to observe the patterns in my yard and to learn something about the plants that I saw spontaneously growing there changed how I view weeds, as well as how I interact with them. Instead of being a source of irritation, weeds are now considered a valuable resource, thanks to the power of observation.

In reference to the aphids that attacked my kale, it did not take a great deal of observation to convince me that I was losing the war. In response, I have adopted a much less aggressive approach that fits better with the natural patterns in my garden, rather than fighting hard against them. Currently, when I notice evidence of aphids, my first strategy is to do nothing. Sometimes, ladybugs will spontaneously appear to knock back the aphid population. If that does not occur, I begin to take action at a leisurely pace, pulling infested plants once they are no longer appealing to harvest and leaving the survivors behind. As the weather warms, I put up shade to cool the garden by a few degrees, which will often extend the life of kale and other leafy greens by a few weeks. Usually, one or two plants will resist infestation and withstand the rising heat long enough to flower and set seed. These plants will become the parents of the next generation of plants. By fostering the survival of the fittest offspring, over time, my garden will become more resistant to aphids and heat without any need for soapy sprays, maximizing the harvest using strategies taken from nature's own playbook.

ACCEPT FEEDBACK

Observation alone can accomplish nothing. After observation comes interaction. Based upon what we learn through our observational findings, we can begin to apply solutions. Then we observe again to discover the effect our applied solutions have had, and to make adjustments accordingly. Observe, act, observe again and adjust: this is the path to sustainability, greater effectiveness and increased efficiency in our lives.

Although it may sound simple to accept feedback and to adjust, it is not always so easy. Humans tend to get stuck in our ways and resist change, sometimes even when our habits beget negative results. And while it may be commendable to pursue dreams and ideals despite opposition, sometimes it's better to change course. When I first got into seed saving and heirloom vegetables, I was very excited about the potential benefits, such as the ability to save the seeds and to grow interesting varieties. I eagerly planted *Black Krim*, *Yellow Taxi*, *Green Zebra* and several other uncommon types of tomatoes. Disappointingly, the plants slowly succumbed to fungal disease, most likely Fusarium wilt, which persists in Phoenix soils. Only a few vines survived to produce tomatoes. The next year, I moved the tomato patch to another area of my yard, with identical results. Determined to achieve success, I tried growing heirloom tomatoes a third season, receiving little harvest despite taking measures to crowd out the fungus with beneficial bacteria. Finally, it dawned on me that I was losing the battle and I changed my tactics. I now grow a mixture of heirloom varieties and hybrid tomatoes that are resistant to Fusarium wilt. Although hybrids are not my first choice, having a few hybrid tomatoes in the garden ensures that I receive an adequate harvest while I continue to make my yard a more hospitable place in which to grow heirlooms.

SMALL AND SLOW SOLUTIONS

Discussion of heirloom tomatoes leads us to the next Permaculture Principle: Using Small and Slow Solutions. When I first noticed the fungal problem associated with hybrid varieties, I employed gentle techniques, cutting off affected vines and spraying the plants with a mixture of baking soda and water. I was hesitant to apply quick remedies, such as commercial organic fungicides, due to their cost and the potential toxicity to bees. I was also hesitant to enter a battle with the fungus that would require repeated applications of fungicide in perpetuity. Instead, I am taking the slow route and working with nature to accomplish the goal and to create a more sustainable outcome. To that end, I rotate the tomato patch annually and amend the soil with effective microbes in an attempt to discourage Fusarium on my property, and I collect seeds from heirloom varieties that thrive despite the presence of the fungus. Over time, the soil should improve, giving my heirloom plants a fighting chance. And since I am saving seeds from the most resistant specimens to propagate the following season, my plants should become more immune to the fungus over time. It may take many seasons to accomplish the goal, and I may never eradicate Fusarium completely. But the benefits of patience will be a stronger, more resilient garden that is healthier, requiring less work and fewer inputs in the long run.

DESIGN FROM PATTERNS TO DETAILS

My sage advice to gardening students has always been to smooth out the earth in their beds prior to planting seeds. I would say to them, "I don't know why seeds prefer flat, smooth surfaces, but they do." In order to smooth the soil, I would recommend using a garden claw to loosen the top few inches of soil, followed by dragging a piece of wood over the terrain to create an even surface. This advice was based on my experiences over the years and anecdotal evidence gathered by talking with other gardeners who expressed a similar observance that seeds seem to sprout better on smoothed ground. Although I had observed and understood the pattern with a small-scale focus on my own garden experiences, my understanding deepened when I began to look for larger patterns in nature.

Interest in food forests is what brought an important natural pattern to light and provided an epiphany about why smoothing the ground seemed to increase seed germination rates. *Food forests* mimic natural forests in that they are perennial and largely self-sustaining, although plant selections are human directed to meet human needs. A mature food forest consists of trees, bushes, vines and low-growing plants that produce food, fiber, pollinator habitat and medicine.

When establishing a food forest, a gardener imitates the natural establishment of a new organic forest. Nature tends to fill empty, barren or disturbed ground, creating forests out of fields, restoring the land following natural disasters, and reclaiming human civilizations when they are abandoned. But forests don't appear overnight. They follow a general and predictable pattern of plant succession. Small, annual plants are the first to grow. We call them *weeds*, and they are the hardiest, most aggressive plants on the planet. Gardeners and landscapers are faced with evidence of their resilience when weeds pop up uninvited in the garden, between cracks in the pavement, and in some of the most inhospitable and unexpected places. These rugged plants seem to prefer disturbed soil surfaces, and their root systems are able to push their way through even the most compacted soil, literally breaking ground for the plant life that will succeed them.

Many of the plants that we consider weeds are the ancestors of common garden annuals. Modern carrots and parsley are domesticated forms of wild carrot, aka Queen Anne's Lace, which is considered an invasive plant. Arugula is the salad form of rocket weeds, and many wild varieties of garlic, onions and mustards are closely related to their garden counterparts.

In the natural order, weeds are the first plants to show up on bare ground following a disturbance, such as fire, flood or catastrophic winds. Thus, it seems to make sense that garden annual and biannual plants, relatives of weeds, would thrive following a man-made disturbance, such as tilling, plowing or digging. So, while I am no fan of disturbing the deeper soil profile, which is rarely disturbed in nature, agitating the surface of the soil appears to be beneficial. By imitating nature's pattern for reclaiming bare ground, I can foster better germination in my vegetable garden.

Observing natural patterns and designing our human systems to work within them has the power to save hours of work and an enormous amount of energy. But one of its greatest benefits is to avoid design errors, both large and small. On the large scale, a friend recounted the story of a home that was built in northern Arizona. The house stood in a ravine between two mountains, and a large swath of forest was cut around

the property to create an expansive view. In so doing, breezes that whipped through the crevasse at accelerated speeds were no longer buffered by the trees. The home became an unpleasant environment, constantly buffeted by strong winds. This design flaw was not easy to remedy, and could have been avoided by a simple site analysis during the planning phase that considered natural wind patterns.

On a much smaller scale, observation of natural patterns could have saved me a headache in my landscape. The Micro Farm Project had a very long block wall that was bare, ugly, and blazingly hot in the summertime. I had the great idea to plant heat- and drought-tolerant plants to shade and beautify the wall. The only hitch in the plan was our dog, Pepper, who patrolled the wall all day long. Before I dug the garden bed along the length of the wall, I hadn't noticed that she had a habit of constantly running back and forth right through an area in which I had planted flowers and small shrub starts. The flowers were almost immediately decimated as she dashed through them without concern for my handiwork, a pattern that I realized was going to be nearly impossible to block or break. The first garden was destroyed before I could stop it, but I replanted the shrubs, protecting them temporarily with chicken wire cages. Had I taken time to observe the natural patterns occurring in my yard, I could have avoided the wasted time and expense of planting flowers that would be trampled.

THE PROBLEM IS THE SOLUTION

Every circumstance and resource in the landscape is either an advantage or a disadvantage, depending upon how it is employed. When designing an urban farm, various problems will be encountered. Perhaps there will be obstacles in the terrain, such as a boggy area or pesticides on a neighboring property. Complications may arise in farming systems, such as excessive animal waste or attraction of unwanted wildlife to your farm. When a challenge arises, the urban famer is faced with a choice of options. One option is to eliminate the problem or reduce its effects. This is sometimes the only logical course of action. But, in many cases, problems contain potential blessings when the option is chosen to view them from a different perspective. To find the silver lining, ask *what is the reason for this "problem" and how can I turn it around to become a benefit?*

At The Micro Farm Project, we had a persistent pigeon problem. The birds were attracted to our animals' food, fruit on our mulberry tree, and to eaves on the patio that became prime nesting locations. We first attempted to eliminate the problem by placing plastic owls in various locations, hoping that the presence of a perceived predator would scare them away. The dauntless pigeons were undeterred, and continued to

visit the patio, leaving behind messy, purple waste that stained the concrete and the furniture. To solve the problem, I first considered why it existed. Clearly, we were providing an attractive environment for them on the patio. The only benefit that I could see for the presence of the birds was the potential to put the pigeon waste to work as a fertilizer in the flower garden. We started by blocking the eaves with wire mesh to prevent them from nesting. My plan was then to set a large birdhouse on a post in the flower bed, complete with a feeder, in an attempt to lure the birds to the garden where they would do some good for the flowers. The season changed and the pigeon problem diminished, so I did not erect the birdhouse. But should they return, I have a plan that I hope will transform their presence from obstacle to asset.

WORK WITH NATURE; NOT AGAINST IT

The goal is always to observe what nature does and to mimic it. Anything else amounts to swimming upstream. That is not to say that the natural world calls all the shots. If I left my property completely to natural processes, I wouldn't have a house, a garden, or any other man-made structure. Without human intervention, things get messy. However, if I use my human creativity and powers of observation to design structures and systems that mirror those that occur naturally, a lot less work is required for maintenance.

My biggest mistake in raising livestock animals in the city was overcrowding my property. In nature, animals have room to roam, and populations are controlled via predators and food scarcity. Thus, waste does not get out of control, creating pollution, but is continually recycled as a nutrient source. At The Micro Farm Project, we unwittingly eliminated the natural controls and encouraged population growth. Though this had some advantages, it also had severe drawbacks. A tremendous amount of effort was required to manage the animals' waste. Raking the pens became a daily chore. And while the chickens enjoyed regular additions to their compost pile and humus was being created at a furious rate, it became impossible for me to keep up with the work. Flies proliferated in the spring and fall, no matter how diligent our endeavors were to keep the property clean. Attempts to use natural controls, such as fly predators and allowing the chickens to feed on larvae, were not enough to prevent nor cure the problem.

We had created such artificial conditions on our property that attempting to retrofit a natural control method was futile.

Figure 8.3 shows how our farm began to look like a zoo with cages – which resembled nothing like the natural environment! And the flies became a real problem. I will admit that I resorted to using insecticidal spray to temporarily knock back the flies for a tour of our property. Although the spray was organic, made from marigold extract called pyrethrum, and not harmful to humans, it was definitely a compromise on my part. I knew that the chemical could be harmful to bees, other beneficial insects, and to the general farm ecology. But I had painted myself into a corner by starting off in defiance of nature in the first place. The fix was technically approved as an *organic* treatment by the USDA, which minimized the harmful side effects. But it certainly did not promote the ecological health of our property, and may have actually prolonged the presence of the flies by inadvertently curbing their competitors.

8.3

WORK WITH NATURE

APPLY NATURAL PATTERNS TO YOUR FARM DESIGN

USING ORGANIC MANAGEMENT ON A CONVENTIONALLY DESIGNED FARM IS LIKE PATCHING A TIRE WITH DUCT TAPE

Trying to use organic farming methods on a conventional farm structure is, I believe, the crux of the problem with organic farming and the reason that many people get discouraged with attempts to produce food naturally. Urban farms are often fashioned with little regard to natural systems and processes, and then expected to be maintained via organic and natural means. Impossible!

It is hard to blame the gardeners and farmers for this situation, as many of us start our operations based on commonly accepted designs and techniques. Neat rows of mono-cropped vegetables, herds of animals in pens, and neatly segregated elements are the typical patterns, and the conventional models that we initially followed in the design of The Micro Farm Project. But we quickly learned that these unnatural patterns were difficult to maintain, and that we had unintentionally entered a battle with nature to bring balance back to the ecology of our farm. The flies were an object lesson that we could not ignore. As a natural result of livestock crowding, maggots moved in to decompose the excess waste, eventually maturing into swarms of flies. From our point of view, this was an annoyance. But from the perspective of nature, the larvae were a necessity, an asset performing a service for the benefit of the system.

The choice was ours to continue an endless battle against the natural course of things, or to bring our farm into alignment with it. But retrofitting a farm is much more difficult than designing an ecologically sound system from the get-go. Using the Permaculture Principles as design tools, it is possible to create an urban farm that works hand-in-glove with natural processes instead of trying to fit organic farming methods onto a conventional farming grid.

In this chapter, we will explore many gentle techniques to protect animals and plants from pests and diseases in the landscape. But we must always keep in mind that good ecological design is the first line of defense. By implementing what we have already learned about integrated elements, healthy soil, biodiversity, and the Permaculture Principles, urban farmers can create systems that are resilient, requiring little human intervention in the form of pest and disease control. Therefore, our focus will be on creating resilient systems and preventing problems in the garden, as well as specific techniques to address complications as they arise. We will discover natural pest and disease control methods that will create healthier plants and easier gardening in the long run.

TO TREAT OR NOT TO TREAT, THAT IS THE QUESTION

Producing food can be challenging, and there are many forces that seem to act against the farmer in the pursuit of growing a garden and raising animals. Pests and diseases rank amongst the top complaints of gardeners, and an infestation of any kind can be extremely discouraging. The first line of defense against pests and diseases is healthy soil. Well-nourished plants have a natural ability to resist pests and disease. Feeding the soil regularly with organic matter keeps disease organisms at bay by producing stronger plants. And it also minimizes disease problems another way – by bolstering beneficial soil organisms that crowd out and ward off pathogens. Pests and disease organisms will always be present in the garden. Thankfully, just as a healthy immune system helps humans to ward off ever-present germs in our environment, healthy soil can halt plant problems from emerging or curb their severity.

Even so, issues will sometimes arise. When a problem is spotted, it is natural for our first response to be a desire to immediately and thoroughly eradicate the problem. But this response is complicated by three factors. First, it is not always a simple task to *identify* the problem. Particularly in the case of diseases, many of them have symptoms that look very similar. Incorrect diagnosis can lead to the wrong, or counterproductive, treatment.

The second factor is that the cost to treat an infestation often outweighs the benefits. Pests and diseases do not generally obliterate healthy plants, but move in when plants are already weakened. Treating a plant that was potentially already under-producing or dying prior to contracting the disease may not be a wise use of resources. For example, aphids typically move into cold season crops, such as lettuces and kale, as the weather begins to warm up. It is almost the end of the growing season for these crops, and the aphids' job is to break the plants down so that something else can grow in their place. Treating for aphids may not extend the life of the plants significantly enough to justify the cost or the effort.

The third factor is that *total annihilation* can result in secondary issues. Not only are chemical residues considered to be unhealthy for human consumption, but utterly removing a pest disrupts natural balances in the garden. In the case of aphids, eradicating them discourages beneficial predatory insects that feed on them, and they may abandon the area. As a result, other pest species may proliferate, unchecked by the

predators that previously resided in the garden. Thus, more treatments become necessary, increasing both the cost and the effort required to garden.

For these three reasons, I am purposefully slow and calculated in my responses to the signs of garden infestations. Here are the steps that I recommend when considering how to respond to garden problems:

1. Do Nothing. When damage is noted, inspect to see whether the pest has already moved on. Perhaps the plant has engaged its own defenses and is already starting to heal, showing signs of new growth. Perhaps the injury is minimal, requiring no action. Or, it may be that the plant is at the end of its life cycle and is not worth treating.

   Sometimes, it is beneficial to refrain from intervening in order to let nature move in to create a more sustainable control. As an example, over the course of several seasons, I battled grapevine skeletonizer caterpillars that would denude my grapevine, rendering them leafless. I tired of fighting them, and allowed nature to take its course. The following season, although I saw signs of the caterpillars, it seemed that my grapes had developed their own resistance to them and the damage was so minimal that I did not need to treat. Other gardeners have confirmed this phenomenon, recounting stories of similar experiences. I am not sure how the grapes are defending themselves, but I am very glad that I sacrificed one season's harvest in order to allow nature to fight the battle for me.
2. Apply Manual Controls. When planting, surround tender vegetables with marigolds, basil, onions, garlic and other strongly scented plants that deter pests. If you can see a pest or if disease damage is localized and minimal, cut it off, pick it off, or spray it off with water. Use physical barriers, such as tulle netting, fences or row covers.
3. Use Gentle Homemade or Commercial Treatments. If manual controls are not sufficient and the plant is worth saving, use homemade and organic treatments. Start with the gentlest, such as a soap spray. If that is not effective, move to stronger organic controls. We will learn more about these treatments later in the chapter. It is important to keep the big picture of your farm's ecological health in mind, even when employing organically approved treatments, as they can have unwanted side-effects (such as harming bees and other beneficial insects).
4. Rethink Your System. If none of the above are effective, perhaps there is the presence of a design flaw. Maybe the plant is not well adapted to the climate, and a different variety would fare better. Maybe it was planted at the wrong time of year, was not spaced appropriately, was watered improperly, or was planted in soil

that needs improvement. It can be difficult sometimes to pinpoint the problem. Investigate by talking with other gardeners, reading gardening books, and going back to Permaculture Principle 1: Observation. One season, I attempted to grow a few stalks of corn in the garden. The ears did not develop properly, showing very sparse kernels. I noticed the presence of ants, and assumed that they were the root of the problem. In talking with other gardeners, I learned that ants were not the issue. I simply had not planted enough corn to achieve good pollination. Every strand of silk leads to one potential corn kernel, and each strand must be pollinated individually in order for the kernel to develop. Corn is pollinated by its neighbors via breezes that blow through the patch. The more stalks in a patch, the better the pollination. The next season, I planted a 6 sq ft patch, which produced beautifully, despite the continued company of ants.

QR 8.1

The link in Figure QR 8.1 provides resources to help you with diagnosing and controlling garden problems.

CREATING A RESILIENT SYSTEM

ORGANIC GARDENING STARTS WITH NON-TOXIC MATERIALS AND WOOD SEALERS

Throughout the pages of this book, we have discovered multiple gardening techniques and the health benefits of controlling the chemicals that come in contact with our food. Many gardeners opt to grow plants in raised beds or containers in lieu of traditional in-ground row gardens due to space, toxicity or labor considerations. Due to constant contact with water and weather, those that are made of wood or other porous media often require a sealant to prevent quick decay of the materials. However, many conventional wood sealers on the commercial market are made with solvents and other ingredients that may release hazardous chemicals and volatile organic compounds (VOCs) into the air and your garden soil. Hazardous chemicals may find their way into the plants themselves, and ultimately into your body if you are growing edibles.

I came across some free wooden wine crates that were the perfect size for a small herb garden. The wood was completely unfinished and required a sealant. I began to research organic and natural options, and found many suggestions online, ranging from low VOC commercial products to homemade organic preparations. These products varied widely in cost, effectiveness and ease of use. The following will describe five options that I consider to be among the best:

1. Raw linseed oil is an eco-safe wood preservative that was commonly used before modern synthetic sealers were created. It is an all-natural product that can be purchased in organic forms. It is fairly inexpensive and easily applied with a brush. One drawback is that the oil is very slow drying, and will take days or even weeks to thoroughly dry. If you choose to use linseed oil, be certain to purchase it in raw form, not boiled, which contains additives that are potentially toxic.
2. Soy-based water proofers and wood sealers are non-toxic commercial wood sealers. I looked at several brands. According to their labels, they are VOC free and claim to provide a twenty-four-month seal. They generally contain oil, water and other ingredients. Though soy-based sealers are among the least expensive natural commercial products on the market, I was unable to verify the exact contents of the "other ingredients".

   The numbers cited on a low VOC paint can are measurements taken before any additives are or pigments are added, both of which can contribute to higher VOC level than indicated. And paints can be labeled "VOC Free" if they contain less than 5g/L [How Stuff Works, a Division of InfoSpace Holdings LLC (2008) How Low-VOC Paint Works. Available at: http://home.howstuffworks.com/home-improvement/construction/materials/low-voc-paint.htm. (accessed October 2016).]

   Because of the way that paints are labeled, I was unable to verify whether or not the soy-based products are truly VOC free, or may contain very small levels of these toxins. The individual gardener must weigh the cost and effectiveness of the product they choose, understanding the possibility of a minute amount of toxicity.
3. Walnut oil sealers coupled with canuba wax or beeswax are safe for wood that comes in contact with foods, containing no solvents and no VOCs. Walnut oil is easy to apply with a soft cloth. A second application can be made in thirty minutes, and the finish dries overnight. When walnut oil is exposed to oxygen, it polymerizes or *cures,* generally rendering it safe for people who have nut allergies. Commercial walnut oil sealers can be found on the market, or they can be

made by simply applying two coats of walnut oil, topped with a coating of wax once the oil has cured for fifteen to thirty days. The only drawback to this product that I can detect is cost, which is the highest amongst the recommended products.

4. Milk paint is an organic preparation that gives a whitewash finish to wood. Before paint was manufactured and sold commercially, it was made at home with simple ingredients and techniques passed down through the generations. Due to the abundance of milk in early rural America, a combination of milk paint and iron oxide (rust) or animal blood was often used to paint idyllic red barns that dot the countryside to this day, a testament to the durability of the finish.

   As the name suggests, milk paint is formulated using curdled milk or curd cheese, lime and pigment, if color is desired. I rate milk paint high on the list due to its simplicity and ease of use. It can be made with organic ingredients very inexpensively. The paint is time tested and lovely. The only drawback to consider is that milk paint can water spot, a likely occurrence in the garden. After painting with milk paint, I recommend rubbing the dried surface with linseed or vegetable oil. Water spots can easily be removed with a soft cloth and oil. Some consider these spots desirable, adding to the rustic appearance that milk paint provides. A milk paint recipe is posted at the link in Figure QR 8.2.

5. My top pick is homemade wood conditioner, made with beeswax and oil. Jojoba oil is an excellent choice, but mineral oil would also work well in this recipe. Melt and mix four parts oil to one part beeswax. Follow the link or the QR code for the full recipe.

   Beeswax wood conditioner is very easy and inexpensive to make. It provides a beautiful, shiny finish that brings out the natural beauty of wood. The wax protects wood surfaces by repelling water and dirt. It can be used as a wood conditioner or polish, or as a sealant applied over a base finish of paint. Figure 8.4 provides basic instructions for making your own.

   The resulting product is food-safe, non-toxic, and you don't have to wear gloves during the application process because it is good for your skin! Additionally, when applied to children's furniture or toys, it is edible and won't harm your little ones if they chew on the finish.

8.4

WOOD CONDITIONER

1 PART BEESWAX PASTILLES
4 PARTS JOJOBA OR MINERAL OIL

- **MELT WAX IN A DOUBLE-BOILER OR SLOW COOKER ON LOW HEAT**
- **WHISK IN THE OIL, STIRRING UNTIL THE MIXTURE IS SMOOTH**
- **REMOVE FROM HEAT. STIR TO PREVENT SEPARATION OF WAX AND OIL AS IT COOLS**
- **USE A BRUSH TO APPLY 2 COATS TO WOOD. LET THE 1ST COAT DRY BEFORE APPLYING THE 2ND COAT**

Apply this wax to your wooden garden beds or containers as a natural finish, or use it to seal milk paint finishes. Keep a jar of this wax handy to condition your cutting board, wooden salad bowls or butcher block counter tops. Apply it to chapped skin in the wintertime. It also makes a fantastic lip balm. For a large project, this recipe is easily doubled.

Is Pallet Wood Safe for Gardening?

In addition to ensuring that wood sealers are safe for gardening, it is also important to select the right wood that will be used to build the garden. It has recently become very popular to recycle wooden pallets for gardening projects. At The Micro Farm Project, we joined the bandwagon and created a vertical lettuce garden that doubled as a fence. The garden was clever and cute, but I stopped growing edibles in it when I learned that old pallets may contain pesticide chemicals that are problematic for gardening. One of these chemical treatments is methyl bromide fumigation. Methyl bromide is a powerful pesticide that is used to kill invasive species, such as insects and rodents. It has been linked to human health problems as well as ozone depletion. In 1987, the Montreal Protocol aimed to reduce and eventually phase out the use of methyl bromide. This type of treatment is currently banned in Canada and many countries because it poses a health risk to workers handling the pallets.

("Methyl Bromide". EPA. Environmental Protection Agency, 18 Mar. 2016. Web. 27 Oct. 2016. <https://www.epa.gov/ods-phaseout/methyl-bromide>.) Read more at the link in Figure QR 8.3.

In recent years, new EPA standards have led many U.S. companies to use heat treatment rather than methyl bromide fumigation to treat pallet wood. Additionally, the International Plant Protection Convention now requires pallets to display an IPPC logo, depicted in Figure 8.5, which certifies that the pallet was either heat-treated or fumigated with methyl bromide.

The IPPC logo includes a two-letter country code (such as US for the United States), a unique manufacturer number assigned by the National Plant Protection Organization (NPPO), HT for heat treatment or MB for methyl bromide, and DB to signify debarked.

Pallets that are heat treated, kiln dried or debarked are most likely safe for use. Avoid using pallets treated with methyl bromide or pallets that do not display an IPPC logo, or use them outdoors for ornamental gardening only. Read more at 1001pallets.com/pallet-safety.

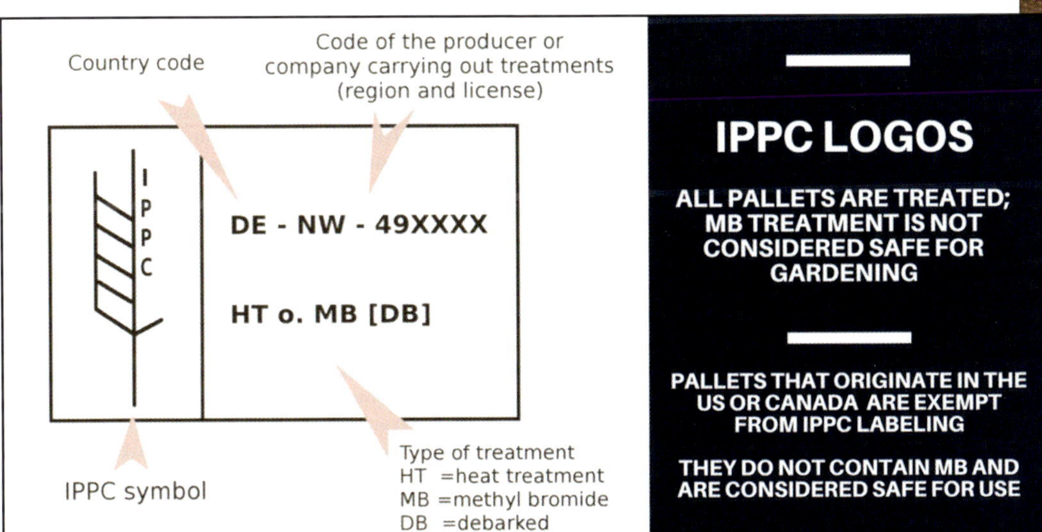

8.5

PLANTING STRATEGIES FOR CREATING A RESILIENT SYSTEM

Many pests and diseases can be deterred by employing a variety of planting strategies. How the garden is planted and maintained has a definite effect on its resulting health, and the following techniques can give your plants a leg-up.

CROP ROTATION

Growing the same plants in the same locations year after year invites pests and diseases, so it is important to rotate the varieties and sites of plantings. Changing the locations of crops deters garden problems in two important ways. First, various plants use soil nutrients in different ways, and rotating plantings prevents soil nutrient depletion. Secondly, pathogens and pests can linger in the garden for a long period of time, perhaps indefinitely if their favorite plants remain. Displacing plantings that serve as their source of food or shelter discourages unwanted organisms, and also moves the crops out of harm's way.

At The Micro Farm Project, we have had problems with Fusarium wilt in our tomato plantings. Fusarium spores can survive for several years in the soil, so when the tell-tale signs are spotted, we refrain from planting tomatoes in that area for at least two years. Phoenix is the perfect area for Fusarium to develop, since dry weather and low soil moisture encourage it. Although types of Fusarium fungi will attack plants other than tomatoes, we obviously have a strain that prefers tomatoes. We discourage it by removing its favorite food source to a new location.

A really easy crop rotation strategy is the *Bean–Green–Fruit–Root Method.* To put this method into practice, begin by growing beans or other legumes in a garden bed during the warm season. Legumes form symbiotic relationships with rhizobia bacteria that are able to take nitrogen out of the atmosphere and make it available to legume roots. The bacteria live in small growths on legume roots called nodules. So when the beans are done producing pods, turn the plants into the soil. As the bacteria dies and the plants decay, plant-available nitrogen will be released into the soil.

Next, plant greens and cole crops during the cool season in the location in which the beans had been. Lettuces, cabbages, kale, broccoli and other leafy vegetables require large amounts of nitrogen for good production. The nitrogen left in the soil by the legumes supports this leafy green growth. One of the functions of nitrogen is increasing chlorophyll production by creating bigger leaf structures with larger surface areas

for the photosynthesizing pigment. Thus, nitrogen fuels fast foliage growth and large leaves that are desirable for greens and cole crops. Corn and squash are also heavy nitrogen feeders, which are warm season crops that can be planted side-by-side with legumes or immediately following.

The subsequent warm season, plant fruiting crops, such as tomatoes and melons. For flowering and fruiting crops, some nitrogen is necessary for the plants to grow vines. But excess nitrogen can be detrimental, fueling fast foliage growth at the expense of root and flower production. As a result, fruit-set will be diminished. Fruiting crops do well in low-nitrogen soils, depending more on phosphorus for healthy roots and prolific flowering, which leads to heavy fruit production.

Finally, replace fruiting crops with root crops in the cool season. At this stage, nitrogen levels in the soil will likely be low. This is a good situation for root vegetables, which are light feeders, requiring little nitrogen to be present in the soil. In fact, if too much nitrogen is present, root crops such as carrots may grow all tops and no roots, or have hairy, misshapen or stunted growth.

Some gardening experts reverse the fruit-root sequence, preferring to plant light-feeding roots following heavy-feeding greens. Others simply prefer to know the family to which their crops belong in order to make certain that their new plantings belong to a different family than the previous one.

There are many ways to perform crop rotation. And even though most rotation methods seem simple on paper, gardening is not always as cut-and-dried in practice. In my garden beds, I will often have a tomato plant survive the winter, remaining in the same place through two warm season plantings. And many plants reseed themselves, popping up of their own accord wherever they want to grow. But following crop rotation principles, even imperfectly, can significantly improve the health of the garden and decrease the number of issues that may arise.

RESISTANT VARIETIES

Plants can suffer from infections caused by bacteria, viruses, fungi, nematodes, and other pathogens. In defense, some plants develop genetic traits that naturally prevent, or greatly inhibit, contamination by these pathogens. Plant breeders carefully select and combine plants that exhibit disease resistant traits in order to produce new cultivated plant varieties, i.e. *cultivars*, that are genetically resistant to diseases. If you notice a pattern of disease symptoms that halt production repeatedly, like I did with my heirloom tomatoes, growing resistant hybrids helps to secure a harvest while the pathogen is present and to discourage the pathogen by removing its food source. Plant labels contain codes that tell if the hybrid is disease resistant. Learn what those codes mean at www.johnnyseeds.com/growers-library/disease-resistance-codes.html.

Another tool in the gardener's arsenal is seed-saving. By collecting seeds from the strongest plants and planting them the following growing season, it is possible to foster increasing resistance to pests and diseases in subsequent generations.

COMPANION PLANTING

Companion planting, aka *intercropping*, is cultivating two or more plant species in close proximity for the benefit of one or all specimens. Variety in the garden has many benefits, from providing shade to shorter plants and soil, to adding nutrients to the soil, and making the most of garden space. It can also go a long way towards curbing the activity of pest species. By intercropping a variety of different herbs, flowers, and groundcovers, some of the plants will deter pest insects, a few will serve as pest decoys, and others will attract beneficial predators.

Insects employ sight, taste and smell, using sensitive receptors on their feet and mouthparts to detect certain crops, sometimes from a far distance. Herbs can repel them by producing odors that either confuse pests or mask the smell of the target plant. Some also emit toxins. For example, marigolds can help control parasitic nematodes (small, microscopic worms that attack the roots of many vegetables) by producing chemicals that are toxic to them.

Marigolds can also serve as trap plants. Trap plants are the favorite food of certain pests, and they are strategically planted to serve as decoys to isolate injurious bugs. Not only do they help to keep the pests away from the rest of the garden, but they also make it easier to find and manually remove unwanted insects. Marigolds are particularly attractive to spider mites, a common pest of tomatoes and other vegetables. Infested flowers can be cut and removed from the garden, allowing the marigolds to continue growing while keeping spider mite populations in check. Nasturtium flowers attract black aphids, and scented geraniums are particularly attractive to Japanese beetles. Use these and other flowers to draw pests away from your vegetables.

Another strategy is to plant a few extra rows of vegetables. Most insects stay on or near the host plants from which they hatch. By growing a few additional rows, infested plants can serve as sacrificial specimens, ensuring the health of the rest of

your garden. Of course, this strategy requires vigilance on the part of the gardener who must keep a close eye on the host plants to ensure that pests are not spreading to other plants. Once pests have used up a particular plant, they may move on to fresh crops. For this reason, particularly damaged or infested plants should be removed from the garden, along with the pests that inhabit them.

In addition to repelling and trapping garden pests, companion plants provide food and habitats that attract and retain beneficial insects that prey on their undesirable counterparts. Plants with small flowers, such as Queen Anne's Lace and sweet alyssum, are favorite nursery plants for green lacewings and ladybugs, both of which are voracious predators of aphids. Bachelor button flowers are a wonderful food source for many beneficial insects, including stingless predatory wasps that hunt a wide variety of prey. Keep your pest population under control by assigning 5 to 10% of your garden or farm area to plants that attract and support beneficial insects. Grow a variety of plants, ensuring that something is in bloom throughout the seasons so that insects will have a constant food source. Native plants are the best attractors for beneficial insects, including bees, as the plants and insects are well-adapted to support each other and the local ecology.

Cultivate a good mix of perennial herbs and flowers in your vegetable beds and along garden edges. Companion planting won't guarantee a totally pest-free garden, but it certainly will help to keep the number of pests and their accompanying damage in check.

INTEGRATED PEST MANAGEMENT

Integrated Pest Management (IPM) is an ecosystem-based approach to pest management that focuses on long-term pest prevention by identifying and disrupting the root causes of pest problems. It uses a combination of tactics and a systematic process to suppress damage caused by pests while preserving the environment and human health.

The IPM process starts with observation. When pest damage is observed, do not panic! A sense of alarm can lead to rushed and possibly erroneous actions. Calm deliberation concerning the problem will assist in making a rational, calculated and more effective response.

It may sound simplistic, but the first step in finding a safe and effective solution to a problem is to identify the pest. Sometimes pests are obvious and it is easy to correctly identify them. Other times, they are difficult to spot. And sometimes, beneficial insects can be confused for pests. Taking the time to inspect and discover what kind of organisms you have is crucial to the process. Once you have identified a problem, try to discover the culprit and where it is coming from in the environment. Ask, what are its patterns? What does it look like in each stage of its life cycle? What does it eat and what does it avoid? Insects and animals vary in their habits and biology, so proper identification and understanding of a pest aids in selecting the most effective controls. For help in identifying organisms in your garden, consult the internet or contact your local Master Gardeners.

Once a pest is correctly identified, conditions within the garden can be altered to make the area less attractive to it. Is the pest hiding under debris? Clean up garden trash and rake back mulch from infested areas. Is it attracted to a dripping faucet or organic fertilizer? Fix the leak and bury the fertilizer. Is it flying into the garden from above? Cover the plants with lightweight row covers. Once the pest's needs and habits are understood, minor adjustments to the environment can be made to significantly deter its activity.

Monitor the garden regularly to spot fresh infestations. Early detection is key to success. Problems that are caught at their beginning stages can be arrested before they get out of control.

The first line of defense is prevention and long-term control as opposed to a reactive, eradication-based approach based on pesticides. Pesticides may be more effective in the short-term for eliminating unwanted insects, but they often come with a rebound effect. Many pesticides are equally effective against pests and beneficial insects, killing them both indiscriminately. Pests tend to have a higher rate of reproduction than beneficials, so their numbers bounce back much faster. Without the presence of predators and parasitic insects, populations of harmful organisms can quickly rise out of control, creating a cycle in which repeated pesticide applications become necessary to keep them in check.

When infestations are present, requiring intervention, IPM employs a variety of tactics, with a focus on the most effective, least toxic and safest methods for the situation. The goal of using multiple *small and slow solutions* builds a wall of defense and prevents pests from evading or developing resistance to any single method. The following are five broad categories of IPM tactics:

1. Cultural methods

Proper cultural techniques go a long way to prevent many diseases. Stressed plants succumb to pests, weeds and disease much more readily than plants that are provided optimal conditions for growth. To reduce the risk and the severity of infection or infestation, choose a location that gives the plant the right amounts of sunlight and shade. Use plants that are well-adapted to the local climate, and avoid watering them too much or too little. Practice healthy soil-building techniques, such as composting. Also, when digging, be careful not to disturb the roots of established plants.

As an example, companion plantings and dense plantings have great value under most conditions. However, if a mildew or other fungal problem appears, it may be necessary to increase air circulation around plants by expanding the spacing between them. Many fungal diseases are precipitated by prolonged stretches of leaf wetness. Good airflow allows foliage to dry more quickly. Also, watering at the soil level will help to keep leaves dry and reduce the humidity that fungal spores require to germinate and spread.

Learn more at RodalesOrganicLife.com/garden/keeping-plants-healthy-heart-organic-gardening and RodalesOrganicLife.com/garden/disease-defense.

2. Physical methods

Physical barriers, such as floating row covers and traps, can intercept pests and hinder their access to host plants. If a pest infestation has already occurred, manually remove them by spraying with a strong jet of water, hand-picking or vacuuming. You can also physically remove their hiding places by mowing tall grasses, removing dead and dying plants, or performing light tillage, depending upon the situation.

3. Biological methods

Biological methods use natural means, such as predators and parasites, in a targeted manner to restrain pest populations. This occurs in a number of ways, starting with protecting and promoting naturally occurring biocontrol organisms by creating conditions in the garden that are favorable to them. For example, when aphids are a problem, a gardener may plant cilantro and dill specifically to attract native green lacewings, whose larvae are voracious consumers of aphids. Predatory insects can also be purchased for release into the garden. However, in my experience, it is better to attract them than to buy them. We once purchased ladybugs. It was great fun to open the package and watch the ladybugs crawl and fly out into the garden. By the next day, however, they were gone without a trace. I suspect that we either did not have enough aphids to support their nutritional needs or that we did not have the right habitat conditions. Thus, the ladybugs abandoned us in search of greener pastures.

QR 8.4

One biological method that we have used effectively is Bacillus thuringiensis, or Bt. The bacillus is a naturally-occurring microorganism that lives in the soil, and which produces a protein that is toxic to many types of larvae. It has been an effective treatment against tomato hornworms and grapeleaf skeletonizer caterpillars that occasionally get out of control on our grapevines, eating all of the green and leaving only the ribs and veins of the leaves behind. Bt breaks down quickly in the environment and is considered to be non-toxic to bees and most predatory insects. Read more about controlling caterpillars at the link in Figure QR 8.4.

4. Chemical methods

As a last resort, a gardener may turn to chemical pest control methods. There are numerous types of pest control chemicals. Some have a wide range of action, having the ability to kill numerous species, aka *broad spectrum* pesticides. Others are more specific, targeting a particular type of pest. Some are highly toxic and persist in the environment for a long period of time. Others are non-toxic and rapidly dissipate.

IPM favors selecting the least toxic, most specific methods of pest control. An example of this would be insect pheromones used together with sticky traps to remove pests from the breeding and feeding population. One application of this technique is a coddling moth trap used to control damage from moth larvae in orchards and garden plants. The traps use a pheromone attractant and adhesive to catch the moths before they are able to lay their eggs. Because pheromones are natural chemicals produced by insects to attract mates, they are non-toxic.

Pesticides sometimes employ natural chemicals that exist in the environment, such as pyrethrins that are produced by plants and spinosad, which is a fermentation product produced by bacteria. While these products are less toxic to humans and are non-persistent in the environment, they are broad spectrum control agents that may damage predators and pollinators as well as pests. Once these products dry, they quickly become inert, having little residual effect. Therefore, when using them, visually identify the specific pest and target it with the smallest application area possible. In other words, hit the pest with the chemical at close range. This will ensure destruction of the pest with as little damage as possible to other organisms in the garden.

Another naturally occurring pesticide is silica dioxide, found in food-grade diatomaceous earth or *D.E.* This pesticide powder is made from the fossilized remains of prehistoric diatoms, and it is completely non-toxic to humans. Diatomaceous earth is effective against soft-bodied insects, particularly insect larvae, as it causes them to dry up and die. Although it feels soft to human touch, it has very small sharp edges that are abrasive to insect exoskeletons. Whenever I clean a goat pen or the chicken coop, I use a sieve or a sifter to spread food-grade D.E. on the ground to control fly larvae. When spreading D.E., be careful not to breathe the dust as it can be irritating to the lungs. D.E. only remains effective as long as it is dry, so regular applications are necessary when pests are active.

One final note about chemical preparations – the word *natural* on a label does not always mean that the product is safe. It is sometimes mistakenly assumed that products marketed as *natural* or *organic* are less toxic than synthetic chemicals that are produced by humans using chemical reactions. This is not accurate, since there are many extremely toxic poisons that come from naturally existing organisms. One example is nicotine sulfate, which is derived from tobacco plants. Although it is a natural chemical, buyer beware. Nicotine sulfate is highly toxic to warm-blooded animals and honeybees. Whenever a chemical preparation is considered, a little research is advisable in order to make the best choices for your farm.

PANTRY SOLUTIONS

There are many home remedies for pest and disease problems that are inexpensive, convenient, non-toxic and effective. These are a few that I recommend:

INSECTICIDAL SOAP: Mix one quart of water with a teaspoon of cooking oil (any kind) and a teaspoon of dishwashing soap. Do not use soap with any type of citrus scent, as it may be damaging to plants. Place the mixture in a spray bottle and go on a hunt for aphids, whiteflies and other pests. When you spot them, spray them directly with the soapy solution. Search on top and under the leaves for eggs, as well.

QR 8.5

The soap will only be effective if it makes contact with the pest, so aim for a direct hit. Then, gently rinse the soap and the dead bugs off plants with plain water.

SOAPY SPRAY BOOSTER: Enhance the effectiveness of your homemade insecticidal spray and deter pests longer by adding something spicy to the mixture. My preferred method is to put a pinch of cayenne pepper with discarded trimmings of onions, peppers and garlic in the food processor with a half-cup of water. Whirl the mixture around. Strain off and discard the trimmings, reserving the liquid. Add the spicy liquid to the soapy spray. When you use this beefed-up version of the spray, the spicy odor that is left behind will serve as a repellent for a day or two. Read more tips concerning homemade concoctions at the link in Figure QR 8.5.

MILDEW MILK: When mildew is spotted, cut off and discard the infected areas, if you can. Be sure to clean your garden tools afterwards to prevent spreading the mildew spores to other plants in your garden. If mildew is spotted on stems and major branches that can't be cut away, spray the area with a solution of 30% milk and 70% water. Any kind of dairy milk will do. Be sure to hit both the tops and the undersides of leaves. This remedy seems to work best if performed on a sunny day as it is surmised that the milk proteins react with the sun, creating a temporary disinfectant effect.

ANT BAIT: Ants love cornmeal, but they can't digest it. When they eat cornmeal, it essentially stops them up fatally. Spread cornmeal liberally around ant-infested areas. Repeated applications may be necessary to eradicate the colony. Keep in mind that ants are beneficial in reasonable numbers in the garden, so consider carefully whether or not the problem is bad enough to treat. If ants are biting you while you work, are damaging plants or appear to be harboring aphids, this treatment will knock their activity back.

FUNGAL SPRAY: Baking soda is detrimental to fungus of all kinds, including powdery mildew, blights and other pathogenic fungi. Mix one tablespoon of baking soda, one tablespoon of cooking oil, one teaspoon of dish soap in a gallon of water. After removing as much of the infected area as you can, spray the rest of the plant with the solution. Do not rinse off. Repeat one or twice a week, being careful not to overdo it

as the mixture can dry out the plants. Plants tend to respond to this treatment better if they are well-irrigated, and a little extra moisture in the soil will help to dissipate the baking soda to limit damage to beneficial soil fungi.

WHAT ABOUT WEEDS?

Weeds can be one of the most persistent and challenging *pests* in the garden. Not only are they unsightly, they take up precious space, compete with vegetables for resources, and some play host to insect-transmitted viruses. Although weeds can sometimes be used for medicinal, culinary or crafting purposes, they are most often a nuisance. Here are a few tips to control weeds in your garden without using toxic herbicides:

1. Clear garden beds of weed seeds prior to planting. This can be accomplished by encouraging weeds to sprout temporarily so that they can be discovered and eradicated. Two weeks prior to planting, water the garden bed to see what pops up. If the outdoor temperature is too cold for germination, cover the bed with black or clear plastic to create a greenhouse. Then, when you are ready to plant, remove the plastic and hoe off any weeds that have sprouted. Now you have a clear, weed-free bed for planting. Note: never allow weeds to grow to maturity in your garden beds. Once they flower, they will spread more seeds in your garden. So, be certain to remove them when they are small.
2. Mulch the soil. Many weed seeds need light to germinate. Covering bare soil with a blanket of mulch blocks the sun and discourages weeds from growing. Apply a 4in layer of coarse mulch, such as wood chips, or a 2in layer of fine mulch, such as shredded leaves. Keep the mulch a couple of inches away from plant stems to prevent disease.
3. Make a homemade weed killer. There are many recipes and suggestions for making your own herbicide. One that has worked well for me is a mixture of vinegar, table salt and dish soap. To one gallon of vinegar, add a cup of salt and a tablespoon of soap. Use a sprayer to apply this mixture at full strength to the top few inches of weed growth before flowers or seed heads appear. Avoid spraying on vegetable plants or on to the soil. In a day or two, weeds will begin to die back. This technique does not kill weed roots, so some weeds may grow back after a period of time. If that is the case, respray, or dig the weeds out by the roots with a trowel.

STRATEGIES FOR LIVESTOCK AND POULTRY

FEED

Many people who raise urban livestock are motivated to cultivate meat, milk and eggs as naturally as possible. What the animals eat plays a big role in producing a vigorous flock and healthy products. The commercial feed market has responded by producing organic options, but they tend to be more expensive than their conventional counterparts. Lewis and I joined an organic feed co-op, which knocked the price down a bit. But the cost was still high enough to make it cheaper to buy organic meats than to raise them ourselves. We began to look for other options.

The first solution was to buy individual grains and mix our own feed. I found that the price to purchase 40lb bags of whole grains was more economical than buying ready-made feed. The recipe was simple: one part wheat, one part barley or corn, and one part whole oats. For extra protein, I would throw in one part fingerling fish food. When I could find flax or black sunflower seeds on sale, those would be added for a treat and to boost Omega-3 fatty acids. During the summertime, I would soak the grains overnight before feeding the mixture to the hens in the morning. Soaking boosts digestibility, and it also gave the birds an extra source of hydration during hot, dry weather.

The second solution was to sprout grains and occasionally grow them to a height of several inches for fodder. My chickens enjoyed free-ranging and dining on scraps from the garden. But our property was not large enough for them to find enough forage to offset many of their nutritional needs. And in the summertime, most of the garden scraps were poisonous, primarily tomato vines and other nightshades. Sprouting grains was an easy way to provide fresh, nutrient-packed greens during the lean months by simply placing seeds and water in a sterile dish, changing the water daily, but otherwise leaving them alone. What I loved about sprouting was that, with just a little bit of effort, the birds were able to receive more nutrition from the same amount of unsprouted seeds due to the increased enzyme content in sprouts, which makes the grains more digestible to the birds.

Sprouts can be produced in just a few days, but if you wait a week you can grow them to fodder length of about 4in. This increases the volume and nutritional value of the feed significantly, by just adding water, natural light and time. Fodder looks like a mat of grass, and chickens love to pick at it. I found that goats also ate fodder, but only if I broke it up and mixed it with their regular alfalfa. If I put a fodder mat on the ground, they wouldn't touch it, which I suppose is normal for goats who like to forage for treats at eye level.

My system for growing fodder was pretty simple, consisting of plastic shoe boxes in which I had drilled small holes for drainage. The boxes were seeded by covering the bottom of each container with a single layer of seeds. The grains were watered and drained several times a day until they grew to a height of about 4in. The fodder was then removed from the container and delivered to the hens or the goats. Between sprouting sessions, I washed the containers in the dishwasher on the sanitary cycle. Learn more about it at the link in Figure QR 8.6.

QR 8.6

THE DIRT

on farming

Growing Fodder For Livestock

CITYFARMINGBOOK.COM/GROWING-FODDER/

LIVESTOCK VACCINES

Whether to get vaccinations for your flock or not is a personal choice. USDA Certified Organic standards allow for vaccination against common diseases, but some livestock managers choose to minimize or forego vaccinating. When weighing the benefits and drawbacks of administering shots, it is important to understand what types of diseases are prevalent in your area, how they are spread, and proper management practices to prevent disease. Contact your local extension office for regional information. The following is some general information about vaccines and how we made vaccination decisions at The Micro Farm Project:

Chickens are often vaccinated as chicks for two diseases: Marek's and Coccidiosis.

Marek's disease is a deadly, highly contagious disease that can cause paralysis, blindness and death. It only affects birds (not people.) The disease spreads quickly and can kill up to 80% of an infected flock. Once a flock is infected, there is no cure. There are two ways to prevent the disease. The first is to vaccinate. The second is to quarantine your flock from other chicken flocks, as well as from wild birds. Small, isolated flocks are less prone to infection by Marek's. But in areas in which the disease is rampant, not vaccinating could result in severe loss.

When you purchase chicks, you may be given the option to vaccinate for coccidiosis, which is an infection caused by a single-celled organism that is found in most environments. The organism can be contracted from the environment directly or from the fecal matter of an infected bird. Healthy birds that are exposed to coccidian generally develop immunity over time. But young birds' immature immune systems make them more susceptible to contracting the disease. If you choose to vaccinate, keep in mind that the coccidiosis vaccine administered in the U.S. is not devitalized in any way, and can actually cause the illness that it is intended to prevent. Coccidiosis thrives in moist environments, so the best way to prevent the disease is to keep the brooder and the coop clean and dry. Unvaccinated chicks can also be fed a medicated starter feed that contains amprol to help them resist coccidia while they develop their natural immunity.

Learn more at chickenwhisperermagazine.com/articles/coccidiosis-chickens-transmission-diagnosis-and-treatment and backyardchickens.com/a/the-great-big-giant-mareks-disease-faq.

In our case, neither of these diseases is prevalent in our area. We chose not to vaccinate, nor to use medicated feed. However, when we hosted tours, we did not allow visitors to enter our chicken coop in order to prevent potential cross-contamination from other flocks.

Goats and Sheep are often given the CD/T vaccine to prevent Enterotoxemia and tetanus.

Enterotoxemia is commonly called *Overeating Disease*. It is caused by bacteria that is found in small amounts in most rumens, but if the pH of rumen becomes acidic, it can reproduce rapidly and cause illness in the intestinal tract. The rumen is the largest of an adult goat's four stomach chambers, and it is filled with healthy microorganisms that break down ingested roughage. Feeding too many grains, giving ruminants baked goods, and any sudden change in feed or pasture can cause the rumen to become acidic, depressing healthy bacteria and precipitating Enterotoxemia. To prevent the condition, limit grains and make changes in feed, hay, and pasture slowly to avoid disturbing the rumen.

Overeating Disease is an extremely painful disease. The condition is difficult to reverse, and animals that develop the disease often scream in agony as they decline rapidly, eventually resulting in death. The disease often strikes young animals with immature rumens. Although we prefer to medicate our livestock as little as possible, we were averse to allowing our animals to suffer so severely. We chose to administer the CD/T vaccine to pregnant females one month prior to birth, and to newborns at six, nine and twelve weeks of age.

Read more at the following websites:

- https://fiascofarm.com/goats/medications-2.htm#cdt
- www.tennesseemeatgoats.com/articles2/enterotoxemiaaka.html
- www.sheep101.info/201/vaccinations.html
- http://fiascofarm.com/goats/index.htm#diseases

FLIES

One of the biggest problems with urban livestock is the presence of flies. In backyard environments, flies can by an annoyance to the people living on the farm, as well as to neighbors. Flies are decomposers that feed on feces and decomposing organic materials. Although the adult flies seemed to be most attracted to fresh droppings from our chickens, we often found their larval maggots hiding under layers of waste that built up quickly in our goat and sheep pens. During seasonal fly proliferations, we found ourselves having to clean the pens daily to prevent build-up. Thus, in my experience, the first line of defense against flies is to limit the number of animals on the property. We deterred flies using pest-repellent plants, which worked relatively well in small areas, such as the patio, but didn't seem to decrease the numbers of flies overall that were on the property (read more about it at the link in Figure QR 8.7.) When maggots were discovered, I exposed them and introduced chickens to the area. Once the chickens lost interest in foraging the area, an application of DE helped to dry the ground and to eliminate any maggots that had survived the birds. To control mature flies, we placed traps in areas where they proliferated, hanging them well out of reach of the animals.

QR 8.7

HOW TO KEEP RAW MILK CLEAN AND SAFE

Fresh, raw goat milk is both delicious and healthy, and milking goats is fun! One you get the hang of it, the chore can be easy and even relaxing. But drinking raw milk can be risky if it is not done carefully, using proper hygiene.

My first recommendation to new and prospective dairy goat owners is to purchase stock from a reputable breeder and a CAE-free herd. CAE stands for caprine arthritis encephalitis, a virus that affects goats, but not humans. CAE is spread via infected milk, colostrum, or blood. The viral infection is associated with acute weight loss, mastitis, pneumonia, encephalitis, and most often, with arthritis. Many goats that are infected with CAE show no symptoms, but they can pass the virus on to their offspring. Since dairy goats must be freshened, or give birth, periodically in order to produce milk, CAE infections are bad news for a dairy herd. Reputable goat breeders regularly have their herds lab tested for CAE, and should be able to produce a recent lab certificate verifying clean results.

Once you have brought your dairy goat home, these are the steps to take to ensure clean, healthy milk on your farm. Note that these instructions do not apply to commercial operations and are not intended to be professional recommendations. These instructions come from my observations and experience. Do your own research prior to drinking raw milk.

Instructions:

1. Keep your goat enclosure clean and tidy. Goats poop a lot! And they will often lie down in soiled areas, placing their udders right on the dirty ground. Removing manure daily and providing fresh, dry bedding a couple of times per week will reduce germs and flies, too! Providing raised platforms on which goats can rest will help to keep their udders clean.
2. Keep a clean milking area. The milk stand, floor, tables and other areas should be swept and washed regularly. On our farm, we milk outdoors, and the biggest risk of germs seems to come from birds that sit on our equipment, looking for stray pieces of goat feed. Sweeping up spilled feed and washing off bird poop is a must.
3. Wash your hands with soap. This may seem like a no-brainer, but this step is critical to clean milking. Using soap and hot water, scrub your hands for at least one minute, scratching your fingertips on your palm to clean nails and cuticles. And speaking of nails, I recommend keeping them short. If you must have long nails, wear a fresh pair of disposable or surgical gloves each time you milk.

4. Clean your doe's udder. At The Micro Farm Project, we keep an array of supplies by the milk stand to keep everything very clean. Our routine is as follows:

   When a doe is brought to the milk stand, she is first brushed to remove stray hairs so that they don't fall into the milk pail. Her udder, teats and back legs are sprayed with food grade hydrogen peroxide and wiped with a clean cloth. Anti-bacterial gel is applied to our hands, and milking begins.

   Discard the first squirt of milk. Goat milk is naturally sterile in the udder, but can be contaminated if germs are present in the teat. By discarding the first pull of milk from the teat, germs are flushed and risk of contamination is reduced. Have a small container, referred to as a *strip cup*, handy to facilitate collection and disposal of the first milk.

   Milk into a clean, cold container. We have two stainless steel buckets that we wash in extra-hot, soapy water daily. The buckets stack on top of one another. The bottom bucket is filled with a layer of ice cubes or an ice pack on which the top bucket rests. In this manner, milk is chilled immediately upon collection. This not only reduces the risk of spoilage, but keeps the milk tasting sweet and delicious. Milk that is not immediately cooled has more of a goaty or musky flavor.

   When milking is complete, strain the liquid through a gold mesh coffee filter or milk filter to remove hair and debris. We milk our does one at a time, straining milk into jars in between and storing them in a cooler with ice until we are ready to take the jars inside and place them in the fridge.

   Finally, clean the teats again. I recommend dipping teats in Fight Bac or Hibiclens using a non-return teat dip cup. An internet search will reveal a number of homemade teat dip recipes, but I personally prefer Hibiclens or other CHG (Chlorhexidine gluconate) treatment because it is non-irritating and when it dries, it seems to form a seal on the tip of the udder that prevents infection very effectively. After dipping each teat, excess cleanser is discarded and the goat is allowed to remain on the stand for a minute or two while the Hibiclens treatment dries. I use this time to scratch my does on the head and give them a treat as a reward for allowing me to milk them.

   Hibiclens can be quite expensive, but it is cheaper than treating mastitis and the resulting loss of milk while the doe recovers. I buy it in gallon bottles, dilute it 50% with water at use, and have very good results.

5. Store milk in sterile containers in the refrigerator. When milking is complete, we immediately date the jars with a grease pencil and rush them to the fridge. Milk lasts a week in our fridge easily with no change in quality. On the rare occasion that we don't drink it within two weeks, I use it to make yogurt or cheese.

It may be my imagination, but I have found that if I scrub jars clean of any milk solids or residue prior to sterilization, milk lasts longer. Perhaps germs are able to survive the dishwasher by hiding in the residue.

6. Thoroughly wash milking equipment with hot, soapy water. Commercial detergents are available for purchase, and many home recipes are posted online. I wash and scrub my equipment with very hot, soapy water and air dry it on a rack. Every time I use the dishwasher, I throw my buckets and straining equipment in with the dishes and wash them on the extra hot setting.

   The washcloths that I use to clean udders are thrown in the clothes washer with a load of white clothing, using hot water, detergent and bleach. Each cloth is only used once before it is deposited in the dirty bin. I have a large number of cloths, purchased at the dollar store, so that I can collect a bunch of them before I have to run a load through the wash. I try to use bleach only on rare occasions, since I have a laundry-to-landscape system. Alternatively, you could use disposable paper towels and compost them.

Drinking raw milk has inherent risks, since it is not pasteurized to remove any harmful organisms that may be present. Research recommended hygiene methods to keep your raw milk safe for consumption, and avoid raw milk altogether if you have immune disorder or other health risks. For those of us who are reasonably healthy, drinking clean, raw milk is good for our bodies and so delicious. Learn more and see photos of our milking system at the link in Figure QR 8.8.

CONCLUSION

As we have shown, in order for a farm to be sustainable, it must be healthy, both in the ecological and the biological sense of the word. And the converse is also true, that when we work to improve the health of the farm, it will naturally become more sustainable. Our properties are improved and our families benefit from living in a healthier environment and eating more wholesome foods. In the next chapter, we will explore more ways to create sustainable systems and thriving systems that last.

FEATURED FARM
THE RODMAN FARM

The internet is a wonderful tool for connecting with like-minded folks. I recently asked a question online about Tattler canning lids, and received a wonderful reply from Heather Rodman. Her response piqued my interest in her farm. I discovered that Heather is a resourceful urban farmer and blogger at The Real Leopardstripes. Her blog motto makes me smile, "Making Stuff, and Breaking Stuff, since 1971".

Heather and her family live in the northwestern region of the U.S., in a high desert area. Although the elevation is much greater than ours in Phoenix, the growing conditions sound so similar with low rainfall, heat, clay soils and caliche. To conserve water and sidestep the soil issue, Heather assembled garden beds out of used skid steer tracks that would otherwise have ended up in the landfill.

"We currently have fifteen tread-beds in operation, located on the north edge of the property, in a pretty compact space. They are sheltered somewhat by a 6ft wooden fence, and a row of Yew trees, to the north, and exposed to the sun southward. Using the treads, we're able to start plants sooner than most of our neighbors, due to their heat retention. This year, we even had, and are harvesting from, volunteer tomato plants, something that no one around here has ever seen before! My husband built the fences and trellis frames ... from pallet wood gleaned from the tractor shop."

Heather and her family started out very inexpensively, using recycled materials and creating their own compost. "This didn't come together overnight. The first two years, we were in a very bad way financially, so every compostable kitchen scrap, dead weed, eggshell, or raked leaf went into the treads – we couldn't afford, yet, to get decent soil to fill them. We broke some of the rules for composting: basically just dumping EVERYTHING in, not stirring it up, and letting the elements have at it."

As they were building their farm, the Rodmans avoided using chemicals. "We stopped using weed killers, artificial fertilizers, and all other chemicals

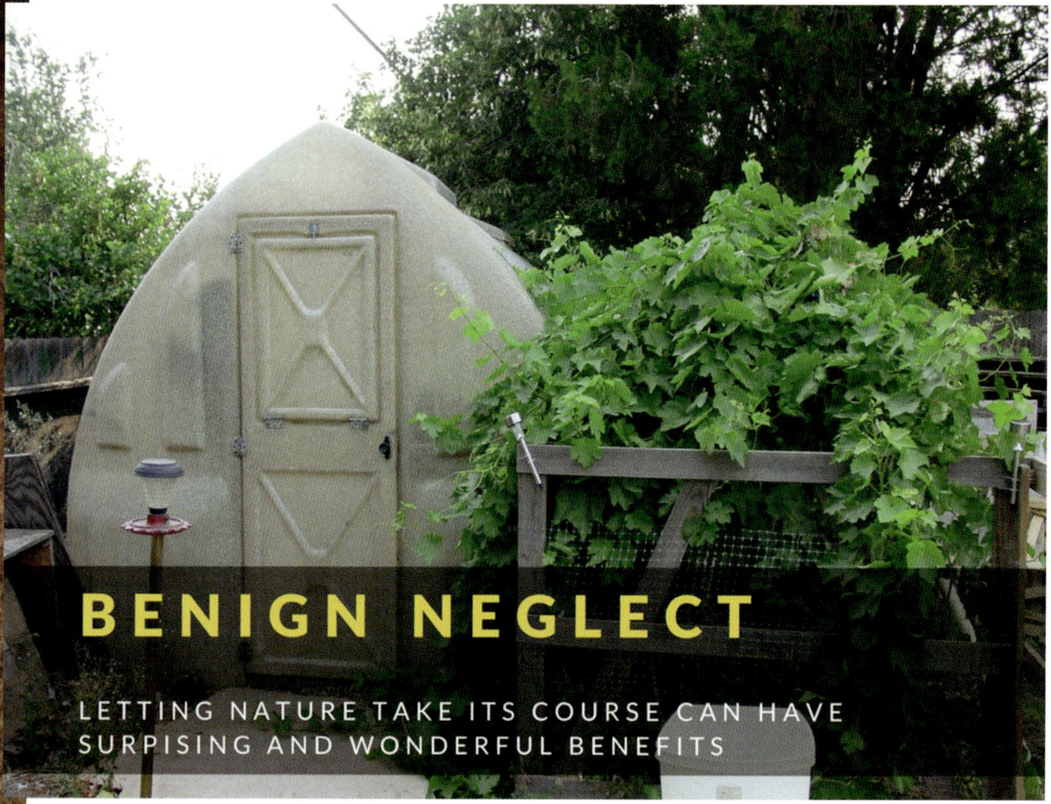

8.6

BENIGN NEGLECT

LETTING NATURE TAKE ITS COURSE CAN HAVE SURPISING AND WONDERFUL BENEFITS

in the yard, as a whole. We gave up on the so-called lawn being perfect, or even watering regularly. The soil was dead, from decades of chemical lawn stuff and pesticides. It'll sound crazy, but I almost cried with joy, the first time I found earthworms in the treads. From there, things began to slowly come back to life." Figure 8.6 shows how lush and green her property has become.

Although Heather has made peace with imperfection that comes with natural gardening, she has been very pleased with the results. "Last year was the first time that we could afford to fill the treads properly, so I laid down cardboard all over, between the treads, to block any thistles or buffalo burrs that might try to spring up. We shoveled finished organic compost into all of the treads, and planted our first crops. We chose tomatoes, peppers, pumpkins, and several herbs. We bought good, tall tomato cages, and devoted all of our outdoor watering budget to just the tread-beds. And were STUNNED when not only did the tomatoes grow up, they grew over, and well beyond, the

treads. The tomato cages collapsed, under the weight … I canned serranos and jalapenos, homemade salsa, and dried large hot peppers by hanging strings of them in every window. But, after canning tomato sauce well into the winter … I finally ran out of box-ripened tomatoes, the last of the crop, in the middle of December. We'll have cases of tomato sauce for at least another couple of years, from that first-year experiment!"

Heather's philosophy is to use what you have and to live and let live. She practices "… a kind of benign neglect. For example, we don't pull all of the weeds – some of them have turned out to be either much-loved by the pollinators, or to be herbs, like the white yarrow that's popped up in the back. (We still don't know where it came from!) This attitude has brought us more joy, and more serendipity, than with any other project we've ever worked on, as a family. It's not the prettiest urban garden, perhaps, but it's the liveliest one in our neighborhood. The variety of pollinators, predatory insects, and surprise flowers and plants has made our growing season an everyday feast of miracles and delights. We do not use pesticides, but we do manually remove squash bugs, and the occasional noxious weed. Most things, even sow bugs and slugs, we just leave where they are – they're just letting us know that we're watering a bit too much. Without a few aphids, you won't have ladybugs. Without some grasshoppers and other leaf-munchers, the mantises have nothing to eat. Everything seems to be serving a purpose out there, even the occasional hornet."

When asked about her future plans for the farm, Heather replies, "We're not sure what we'll be planting next year, but we're firm believers in the intelligence of crop rotation, to reduce pests and diseases. Whatever it is, I'm sure it'll be abundant! (And probably at least three times as much as we planned for!)."

Read the full interview with Heather and see photos of her farm at the link in Figure QR 8.9.

QR 8.9

Chapter

9

Taking the Long View: *Going Beyond Sustainability on the Farm*

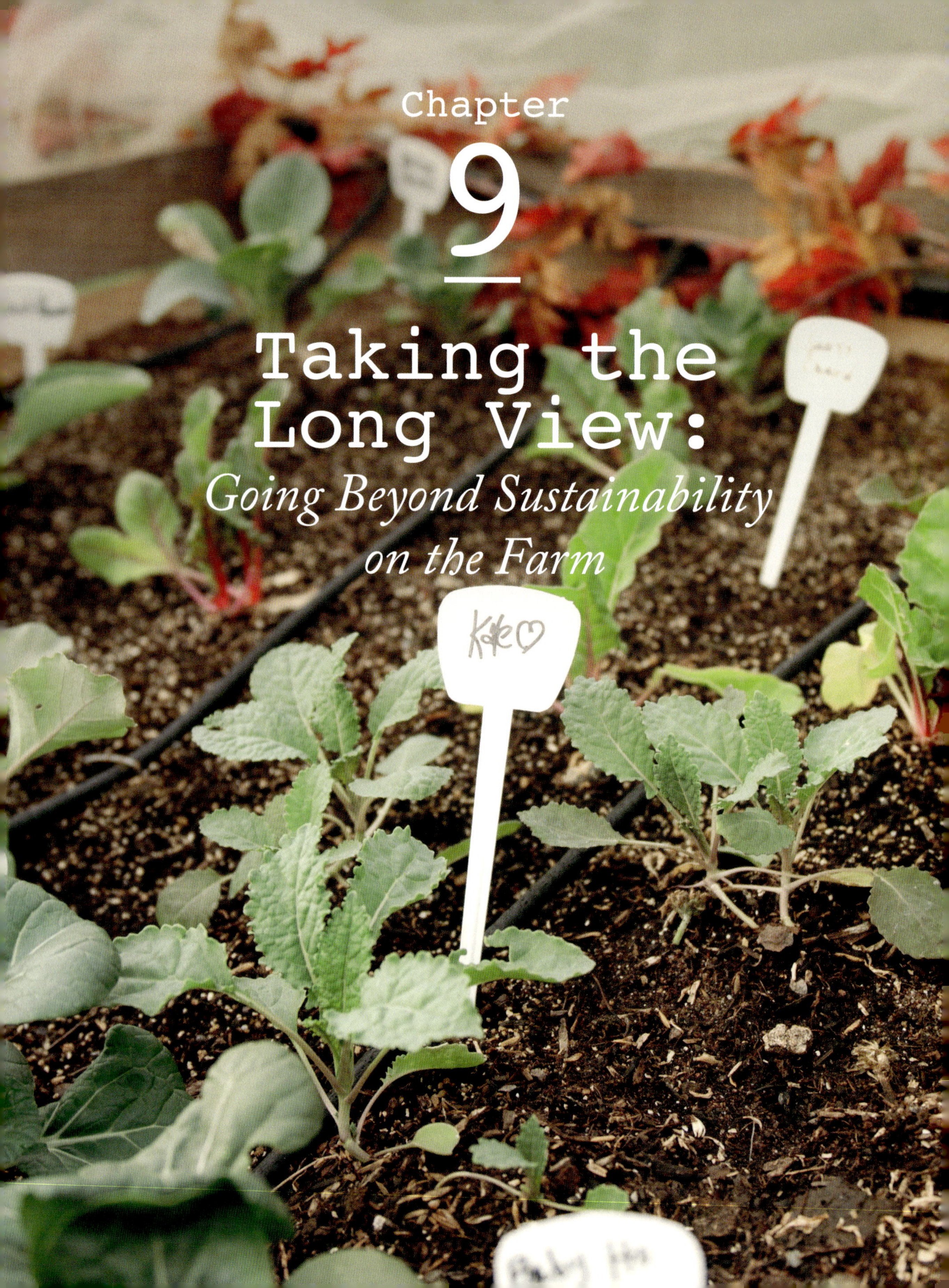

I love urban farming. The smell of damp soil, the gentle clucking of chickens, the sight of a deep orange egg yolk, the act of picking a basket full of colorful produce; each of these calls to mind the sweetness of life. The farm's daily tasks are a meditative practice, marking the weeks with a comforting routine and connection to nature. Maybe it's in my blood, as I come from a line of agriculturalists and gardeners, but I'm happiest with dirty fingernails and a farmer's tan.

As much as I prize our little farm, it is, after all, an avocation, a cherished hobby, but not the primary way that I make a living. I am very fortunate that my vocation as an urban farming educator is related to my hobby, and that I am equally enthused by teaching and writing about farming as I am by actually farming.

The irony is that research often takes me away from the farm. In June 2015, I had the opportunity to visit two amazing operations, Scott and Laura Murray's *Edge of Urban* backyard farm and Jason Mraz's organic avocado ranch, both located in California. Lewis and I decided to take the whole family and make a vacation of it.

That proved challenging, as there was no one left behind to keep the farm running while we were away. I was fortunate to have a trusted friend to oversee the operation and a group of students from the Urban Farming Program at Southwest Institute of Healing Arts who could earn externship hours by milking my goats and feeding the other animals. The farm was managed well. But, as expected, our ornery doe Annie Oakley gave the substitute milkers a hard time, and her milk production plummeted. I had to work diligently upon our return to keep her from losing milk altogether.

In the meantime, other travel opportunities began to arise. But I was hesitant to leave the farm again. The externships were by then completed, and I could not afford to hire help. My children were willing to fill in for me, but as they were teenagers with busy schedules and heavy workloads at school, I did not want to place the extra responsibility on their shoulders.

And thus, in order to fulfill my work responsibilities and my career dreams, we had to redesign the farm to become sustainable under the current circumstances. This meant carefully selecting new homes for our beloved goats, determining not to hatch a brood of new turkeys after Thanksgiving, and creating a garden club whose members could care for the gardens while we were away in exchange for a share in the produce. I had to let go of the costliest aspects of the farm in terms of time and money, at least temporarily, in order to make a living.

This illustrates the crux of sustainable urban farming. When I first heard the term *sustainable agriculture*, I rightly surmised that it had something to do with the environment. It made sense that I could not exhaust the soil and deplete natural resources if I wanted to run a sustainable operation. But I quickly learned that full sustainability goes beyond environmental concerns, taking into consideration time, energy and finances. In other words, a truly sustainable system is ecologically sound and it does not exhaust the farmer, nor deplete his or her economic resources.

Sustainability simply defined is the endurance of a system. In my view, any farm or garden endeavor will only be sustainable, enduring for the long term, if it supports and replenishes three essential pillars of farming: the environment, the farmer's energy and the economic aspects. I call these the 3 Es of Urban Farming. Figure 9.1 shows a diagram of the 3 Es, representing them as mutually supportive spheres that cannot be separated.

9.1

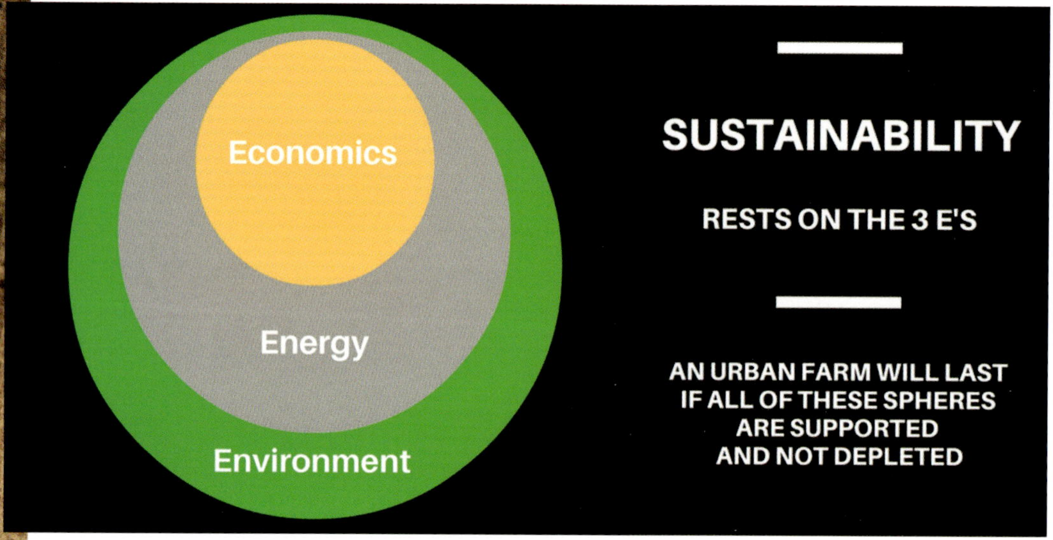

I placed the environment as the largest, all-encompassing sphere because any lasting, productive farm operation rests on a healthy ecology. In previous chapters, we discovered that farm systems can be designed to support and to mimic nature, supporting healthy environments that increase in abundance over time and bolster human well-being. In truth, ecologically sound, regenerative systems are easier and cost less to run than extractive systems, thus supporting the inner spheres of energy and economics.

The middle sphere is energy. Certainly, it is important to use energy wisely and to conserve it, and this includes *farmer-energy*. I define *farmer-energy* as stamina +

enthusiasm. To illustrate, when we started our farm, I was not yet forty years old, young enough to have great stamina for the work. I also had enormous enthusiasm for the new endeavor. Consequently, I was able to transform our property into a farm and maintain it, even while my husband was away on military deployment. I am now approaching fifty years of age, and my stamina has decreased. I am also not as enthusiastic as I once was about certain elements of the farm, and so my motivation and energy level to maintain those systems has decreased. Thus, I have chosen to maintain only those food production systems that yield the most and that I love the most. Meanwhile, I have eliminated systems that sap my energy and have become a drudgery rather than a pleasure to manage. I encourage new farmers to start small, carefully considering the amount of time, energy and ongoing enthusiasm they will have to devote to the project. First, know thyself. Then, design your farm to match your abilities and your temperament.

The smallest sphere is economics. Right now in the U.S., demand is growing for natural, local produce and consumers are becoming more and more willing to pay a premium for it. In response, a number of innovative metro urban farmers are coming up with original ways to meet the demand. Most farmers enter the occupation because they have a passion for growing food and a desire to produce a superior product. Pride and satisfaction in what one produces are certainly important. Coupling these with a decent return goes a long way to preventing burn-out that farmers sometimes experience, which is often associated with poor financial health. The joke is that to make a million dollars in farming, all you have to do is start with 2 million. But, all kidding aside, a farm that is wisely designed and managed will yield satisfactory economic return.

Whether you grow food as a business, or simply to offset your grocery bill, the farm should benefit the farmer financially. At The Micro Farm Project, we have, on occasion, eliminated elements that cost more than they yielded. However, I place economics as the smallest sphere due to the fact that we sometimes maintain projects that have a low economic yield, but benefit our family in other ways. For example, I grow flowers purely for their beauty and we raised sheep for the joy and the benefit of having clean meat for our kids, despite the expense. That being said, a farm that drains the farmer's bank accounts will likely fail. Since we want our farm to last in perpetuity, we are always looking for ways to grow and raise food more economically, and to create extreme abundance on our property. And, certainly, if one's desire is to create an urban farm business, the economic sphere would move closer to the forefront in importance.

While previous chapters have been devoted to organic gardening and animal husbandry methods that preserve the environment and conserve resources, this chapter will also explore the ecological functions of a farm, as well as the business end of farming and *profitability*. Can an urban farm be profitable environmentally and economically, as well as physically and emotionally for the farmer? To begin to answer the question, we will look at the economic functions of a farm and practices that can increase a farm's bottom line. We will also look at the ecological and energy functions of a farm that support economic profitability while enhancing the environment and the farmer's quality of life.

ECOLOGICAL FUNCTIONS OF THE URBAN FARM

On a visit to New York City to investigate urban farming in April 2016, I was astounded and impressed by the city's flourishing agricultural life that did not seem to be at all deterred by the lack of open land. By means of both traditional and innovative methods, residents grow food and ornamental flora in every nook and cranny. Dozens of community gardens are tucked between buildings or are situated in parks, such as the vegetable farm at busy Battery Park. Millions of colorful flowers bloom in tree boxes and containers that line every street. Window boxes and front stoops bear fruit and blossoms. Areas that were once covered with vacant cement now boast

9.2

PLANT BENEFITS

EXPANSIVE PATIO CONVERTED TO A GARDEN IN NYC

**ADD BEAUTY
ATTRACT POLLINATORS
FILTER POLLUTION
SEQUESTER CARBON
COOL HARDSCAPES
INCREASE BIODIVERSITY
PROVIDE FOOD**

expansive container gardens, and even the rooftops are put to use for growing food in small-scale, but intensive, systems. Figure 9.2 shows a thriving garden covering the underutilized patio at Riverpark farm-to-table restaurant in Manhattan.

Flora growing in the city certainly beautifies and softens the otherwise hard landscape, but the ecological benefits of urban gardens go far beyond the aesthetic value. Plants are wonderful filters for toxins, both air and water-borne, an important function for city dwellers who are persistently exposed to pollutants. In contrast to conventional farming methods that include deforestation and mono-cropping, urban agriculture enhances the metropolitan habitat and increases the variety of life, aka *biodiversity*, as gardeners grow a varied array of plants. Diverse food crop varieties contribute to the stability of the food system by increasing the chances for successful adaptive responses to threats, such as pests, disease or climate change.

In relation to climate change, greenery absorbs excess carbon dioxide from the atmosphere, and sequesters the carbon in plant tissues where it is beneficial. Trees and perennial plantings are particularly useful for capturing carbon and converting it via photosynthesis into carbon compounds that fuel growth. Carbon is stored in the roots, trunks, branches and leaves, aka the *biomass*, until being released back into the atmosphere when the plant dies, decays or is digested. In a home garden, dead plant matter (as well as animal waste) can be composted, limiting the amount of carbon that is released into the atmosphere and storing it in the earth where it is immensely useful for soil and plant life. [Global Issues (2014) Why Is Biodiversity Important? Who Cares? Available at www.globalissues.org/article/170/why-is-biodiversity-important-who-cares#WhyisBiodiversityImportant (accessed November 2016).]

The biological processes of plants can also serve a beneficial environmental purpose. A by-product of the photosynthetic process is water. Water molecules are released from plant leaves by means of transpiration. As the water vapor rises into the atmosphere, it tends to condense into clouds. Cloud formation is an especially important function in areas affected by global warming or drought, one that I want to shout to the rooftops in Phoenix where rock landscapes are the norm. With an aim to conserve water in our dry climate, we have limited plants in the terrain. Perhaps that was the wrong approach, contributing to poor air quality, rising temperatures and disappearing monsoon rains. I would personally love to see what would happen if we began to grow more vegetables, trees and ornamental vegetation, using water-wise strategies. Perhaps we could reverse the perpetual drought and create an oasis in the desert. But I digress …

Growers in excessively wet climates may view increased cloud formation as a negative result. However, water-loving trees that are adapted to grow in marshy areas can be extremely useful for improving drainage in waterlogged spaces. Wet soil trees tend to use large amounts of water, essentially sopping it up and removing it from the landscape. Learn more about wet soil trees at GardeningKnowHow.com/ornamental/trees/tgen/wet-soil-trees.htm.

Additionally, biodiversity supports pollinators. An apparent decline in natural and managed pollinators is concerning to farmers and gardeners. Diverse plantings provide habitat to pollinators that uniform monocultures cannot furnish.

Although this is not an exhaustive list of the ecological benefits of urban farming, it is clear to see that the activities of home agriculturists support both the farm microcosm as well as the larger local environment. The amazing fact is that any farmer can improve the ecology of their own property while simultaneously having a positive effect on the greater community. In my estimation, the dual benefit is extremely motivating to consistently apply natural, constructive growing methods, knowing that they are making a beneficial impact, both personally and potentially globally.

ECONOMIC FUNCTIONS OF THE URBAN FARM

Industrial farms operate much like a factory. Production starts with trucking expensive inputs on to the property in the form of fuel, fertilizers, equipment, seed and packaging items. These materials are utilized and sent back out the door as produce. In contrast, the job of a Permaculture farmer is to limit inputs that come from off site, and to make use of readily available natural resources to grow food and generate an income. Throughout the pages of this book, we have seen how a farm can function with very few outside resources. By consciously narrowing the number of inputs that are trucked on to the farm, depending instead on the natural energy of the sun, rainfall, pollinators, atmospheric nitrogen and other free resources, expenses are limited and profits are expanded.

There are many varieties of backyard farms, from hobby farms to commercial enterprises. No matter where a farm falls along the spectrum, economic concerns play an important role in the long-term sustainability of the operation. It is easy to

understand the role of finances if a farm is structured as a business. As for hobby farms, economic benefits may be secondary to the pleasure of gardening and producing one's own food. However, if the cost begins to outweigh the non-monetary benefits, the longevity of the farm will be at risk. This does not mean that a hobby farmer must sell items and receive a profit. There are several economic functions that an urban farm can serve, and *which* function it fulfills depends upon the reason for which the farm exists in the first place.

Home gardens and farms have three main economic functions. The first is to save the farmer money by replacing items that would have been purchased commercially with items that are produced at home. At The Micro Farm Project, propagating onions, garlic, herbs, milk and eggs replaces many trips to the grocery store, and these are the items that benefit us the most monetarily. Read about how I grow onions at the link in Figure QR 9.1. For many urban farmers, trade also offsets the budget. Getting to know other growers with whom one can swap items is a great way to offload excess produce while receiving items that one wants or needs.

QR 9.1

The second economic function is to share the excess and provide sustenance to others. One of the most exciting benefits of gardening is the ability, and sometimes the *necessity*, to give a portion of the harvest away to others. Bumper crops not only offer the opportunity to preserve the excess for the off-season, but also to bless others with a healthy and delicious gift.

The third economic function is to earn an income. Many growers offset the cost of home food production and diversify their income with earnings from the farm. There is an expanding market for fresh eggs, milk, meat and produce from local sources. Laws vary from state to state concerning the sale of backyard goods, so check with your state's department of agriculture concerning the rules in your locale. For instance, in Arizona, one can sell up to 750 dozen eggs without having to purchase a license as long as the producer registers with the state as a Nest Run Egg Producer, displays *Nest Run* (i.e. not graded by USDA standards) on each egg carton, and complies with various other rules.

MAKING A LIVING

Some urban farmers have ambitions to offset a large portion of their budgets or replace their income with farm activities. The prospect of earning a living via farming is challenging, and good production alone will not make a farm profitable. These are a few strategies that small holders can employ to give them a financial edge.

VALUE ADDED: Growers can increase the value of raw products by transforming them into cheese, pickles, jams, flour, salves, soaps and other *value added* commodities. The term *value added* refers to the worth of these changed items, which is often enhanced. To illustrate, during the wintertime in Phoenix, the price of citrus plummets. Many people grow oranges, grapefruit and other types of citrus in their yards. When it ripens, folks who have citrus trees often have much more fruit than they can possibly eat, and it is difficult to offload it as there is so much of it ripening in the valley at that time. During the seasonal glut of fresh fruit, citrus farmers are faced with the prospect of either shipping their produce to colder climates, or transforming it into products that will sell locally and keep for a long time, such as jams and curds.

Another example of applying the concept of *value added* would be to use herbs or calendula, which might sell for a dollar or two per bunch, to create a high dollar salve or tea. Turning milk into cheese, wheat into bread, rhubarb into pie, or any other creative way of adding value to farm produce, can help you to stand out in the market against the competition and maximize income potential. When considering options for value added products, check with your local municipality concerning rules and procedures, particularly for food items that tend to be more heavily regulated than non-edible commodities.

FOOD HUBS: Small producers often have passion and skill to produce high quality commodities, but do not have the financial means to store, market and distribute items to the public. In order to afford these costs, associations are formed by third parties or the farmers themselves to share resources. Generally, how this works is that local and regional producers pay a small fee to participate, and the food hub organization manages the accumulation, storage, distribution, and marketing of the products to individuals, restaurants, retailers, and institutions. Resources may be pooled to provide expensive equipment, such as refrigerated trucks or commercial kitchens. Farmer's markets are a common example of a food hub, but organizations vary according to community needs and desires.

COMMUNITY SUPPORTED AGRICULTURE (CSA): Community Supported Agriculture is a way for producers to raise the capital needed to operate their farm by offering shares in the farm to consumers. It generally works as such: A farmer offers a number of *shares* to the public. The number of shares is based on the projected harvest quantities. Local customers are able to buy a share in the expected harvest, which is basically a membership fee that supports future crops with money up front. In exchange, members receive a box of fresh produce every week during the growing season, picked up directly from the farm or delivered to drop-off locations.

CSA subscriptions are advantageous to small farmers for several good reasons. First, they provide up-front financial resources for starting or expanding a farm, and the members share in both the benefits and the risks. If the farm has a bumper crop year, the members benefit by receiving large shares of produce. However, if the harvest is poor, members understand that they will receive smaller shares, and the farmer is not under the burden of reimbursing any money (as one would with a loan.)

Secondly, CSA reduces labor and administration costs. Subscriptions are usually marketed in the off-season, freeing the grower to focus on crop production during the growing season. Member volunteers will often assist with the marketing and distribution efforts. For example, if a member would like to have a drop-off station near their home, it may be incumbent upon them to volunteer to run it for the season and to recruit a number of other members in order to justify adding the new location. When members are invested both personally and financially, a sense of community and camaraderie can develop; members cheer together when crops are outstanding and share the grief when they fail. When a catastrophic event occurs, such as a debilitating injury to the farmer or an unexpected early frost requiring immediate harvest, members are invested in the success of the farm and many will be willing to help out in an emergency.

Finally, CSA allows the farmer to directly connect with local consumers, and to structure the arrangement to maximize the benefits both to the farm and to its customers. In some communities, members may prefer to visit the farm and load their own CSA baskets. In another region, members would rather have shares delivered to convenient drop-off points near their homes. In densely populated urban areas, farms may be too small to support their own CSA associations, opting to band together to form a multi-farm CSA group. Small farmers are free to creatively structure their groups to meet the needs of the farms and the consumers alike.

NICHE MARKETING: The term *niche marketing* refers to small businesses or farms that produce a specialty product or service for a particular segment of the marketplace. Market niches can meet the needs of specialized industries, specific geographical regions, societal trends, ethnic or religious groups, or any other particular segment of society. Sometimes a niche product can be a variation of a common product. For example, in Phoenix there is a growing demand for raw goat milk. It is very difficult and costly to obtain a license to produce this for sale in Arizona, but one local dairy saw the marketing potential and subsequently converted its operation from pasteurized cow milk to raw goat milk. Crow's Dairy now serves numerous restaurants and grocers, offering Grade A goat milk and cheese with very little competition in the market.

Urban farms are often perfectly suited for niche markets. Lacking land area and capital to produce large commodity crops, small farms have the flexibility to discover and meet the specific needs and specialty markets within their own communities. Profits can also be maximized by offering unique, gourmet or trendy products that are priced higher than commodity items. Lewis and I partnered for turkey processing with our friends, Tina and Bruce of Stone Hoe Garden, who were raising several dozen birds for sale to the local market. During the Thanksgiving holiday season, turkeys are sold in retail stores for less than one dollar per pound. But Tina and Bruce were able to sell their farm-raised, heritage breed turkeys for ten dollars per pound, a significant mark-up from the price of a standard supermarket turkey. Demand was high, and customers began asking to reserve next year's turkey nearly a year in advance. By meeting the niche demand for locally and naturally raised turkeys, Stone Hoe Garden amplified the profit potential, despite the limited space available to raise poultry on their 2-acre property. Explore more about raising turkeys at the link in Figure QR 9.2.

TOURS, CLASSES AND EVENTS: One of the missions of The Micro Farm Project is to inspire other gardeners and urban farmers. To that aim, we open the farm regularly to the public. Twice annually, Lewis and I partner with Greg Peterson of the Urban Farm to host tours. The tours and education received by attendees are free, but we earn a bit of money by collecting free-will honorariums from those who are willing and able to contribute. Greg and I also offer classes related to urban farming using the same donation structure. This system has worked very well for us, bringing in a small income while serving all interested parties, regardless of ability to pay.

Additionally, we host private tours for community college programs and other civic groups that provide a financial stipend per attendee. And when we have a farm project to tackle, such as building new gardens or installing a water harvesting system, we offer a hands-on class to the public. Although these classes do not yield much (if any) income, it's a winning situation both for the farm and for the volunteers, who receive a great educational experience while assisting us with farm labor.

Finally, if we want to learn a new skill or complete a project for which we lack expertise, Lewis and I reach out to form partnerships with others who have the skills that we require. On several occasions, we have invited experts to teach a class at our farm. Our part is to fill the class with students and to collect honorariums for the teacher. In return, we receive free education and labor to complete complex tasks. On two occasions, we partnered with a local chef and our friends at Stone Hoe Garden to host a meat processing and butchering class. We provided the sheep and goats, as well as the venue. The chef provided the labor and education. By doing so, we learned how to better process our animals and received butchering services at no charge. The chef benefited by bringing his kitchen staff, whom he could train without having to use restaurant funds to purchase whole animals for demonstration. We also opened up the class to the public. In exchange for labor, attendees received a close-up demonstration and the opportunity to interact with a well-known local chef.

Tips for hosting visitors at your farm are at the link in `Figure QR 9.3`.

QR 9.3

This leads to our final strategy…

DEVELOP A NETWORK: Networks of people do not form overnight or without effort, but if you prioritize developing strong relationships with your neighbors, customers, volunteers, workers, and other farmers, the energy and input they supply can greatly enhance the success of your farm.

Who should you include in your network? Befriending experienced farmers is a must. Unlike some other occupations, urban farmers do not generally keep *trade secrets* and many are willing to help new farmers to get started. By getting to know a few in your area or field of interest, you will have someone to call upon should you need some information or advice.

Secondly, familiarize yourself with your local university agricultural extension. Every state has a land-grant university with extension offices that exist to help local farmers and gardeners. They offer free and low-cost education and many of them have programs specifically for beginning farmers. By taking advantage of their services, you will not only learn about local farming, but you will also rub shoulders with other farmers and discover farm resources that are available in your area.

Thirdly, if you have livestock, consult with a vet and keep their phone number handy for emergencies, as well as for routine care. Not all vets treat livestock, and there is nothing more stressful than looking for a vet as a medical predicament is occurring.

Finally, nurture and expand your network. Begin to invite people to your farm, offering them education, produce and whatever else you have to barter in exchange for assistance with daily activities and special projects. Find out what skills are available to you within your network of employees and volunteers, so that you know whom to call when you are in need of help. If you are warm, friendly, enthusiastic and generous, you will attract folks who want to be part of your project.

ENERGY AND THE URBAN FARM

My husband jokingly says that he is *ambitiously lazy*. In truth, he is far from slothful, but rather *highly efficient*, able to find the quickest, easiest and most economical way to accomplish any task. This is a highly valuable trait for a farmer due to the fact that food production can require a lot of hard work. Growing food is fantastic, but if you wear out or burn out in the process, the value is lost. In terms of the three Es of Urban Farming, we are always considering ways to produce food in a manner that is ecologically and economically profitable, while expending the smallest amount of energy as possible.

In previous chapters, we learned that good farm design places elements strategically to make urban farming tasks easier. At The Micro Farm Project, we installed water spigots near the areas in which water was needed. This reduced the amount of farmer-energy that was required on a daily basis to perform watering tasks. We also considered how our animals could accomplish certain jobs for us. One strategy that we employed in that regard was to place our compost pile inside the chicken run where the hens could turn and aerate the pile for us.

Together with strategic placement of farm elements, well-maintained equipment and structures also reduce expenditures of farmer-energy. Having the right tools for various tasks, in good condition and within easy reach, can make a big difference in how much mental and physical activity are required to accomplish those tasks. Strong, dependable structures keep animals, equipment and gardens secure, minimizing work and worry on the farm. Smooth, clear pathways as well as working gates and latches, reduce frustrations and make farm traversal untroubled.

When considering the simplest way to perform a task, the question is not always what is easiest at the moment, but what will streamline the task in the future. By way of example, I built a new gate for our chicken run and needed a latch. I happened to have a small sliding bolt latch in the shed that was not being used. I suspected that the latch might be too small and not the right type of closure. But for the sake of convenience at the moment, I installed it against my better judgement. For months we fought with the latch, which was difficult to operate on that particular gate. Although it was a small hassle, going in and out several times per day finally wore me thin enough to make a trip to the hardware store for a more appropriate latch. A little bit of extra effort up front, in the form of purchasing and installing a proper latch, was well worth the simplicity it afforded each time that I would enter the run in the future.

Over the years, I have realized that although I know a lot about urban farming, there is always something new to learn. There are tools existing of which I wasn't aware and methods that farmers employ about which I had no idea. For any task, I have found it to be profitable to do research online and to ask around in order to find out the easiest and most effective way to do it. Oftentimes, I am amazed and delighted by the creative, ingenious and simple things that I learn from other farmers to streamline tasks.

I have discovered that although occasional challenges can be exciting and energizing, the day-to-day tasks of food production are a whole lot more fun when they are easy than when they are hard. Thus, I have adopted my husband's aim to be *ambitiously lazy* on the farm. Every farm and every farmer are different, and it is important to match one's farm design to one's own ambitions and energy level. By cultivating high efficiency, growing food does not wear the farmer out, but energizes and nurtures one's whole being.

THE BUSINESS OF FARMING

Although we have discussed quite a bit throughout the pages of this book concerning the production end of farming, we have not covered the business aspects that contribute to the efficiency and economic health of the urban farm. Personally, I would love to putter around the farm and avoid attending to business paperwork. But just as crop and harvest planning increase yields and efficiency, spending a little bit of time planning your business and tending to administrative tasks brings focus and peace of mind to urban farming.

Here are four administrative tasks to consider:

BUSINESS PLANNING: Whether your intent for starting an urban farm is for-profit, non-profit or strictly for personal benefit, spending the time to prepare a business plan is a key component to the success of your operation. The process of preparing the document forces the farmer to think carefully through the reasons that the farm will exist and its specific goals and mission. Although these can change over time and the business plan is not set in stone, a mission statement and written goals can help the farmer to stay on track with what is really important to the farm, and avoid distractions that will certainly come along. As for public farms, the business plan gives credibility to the operation, articulating clearly what the farm will look like and how it will be sustained financially. Expressed clearly in writing, the document serves to bolster confidence in customers, volunteers, donors and financial institutions as they consider whether to invest their time and money into the enterprise.

Although the prospect of preparing a business plan may seem daunting, there is help available and free resources for farmers. The U.S. Small Business Administration has resources on its website at SBA.gov, as well as offices in every state that provide mentors to entrepreneurs to help them prepare business plans and guide them in other aspects of running a business. Business planning tips and resources are posted at the link in Figure QR 9.4.

BUDGETING: Part of business planning is preparing a budget. Although negative connotations exist around the concept of budgeting, it is a valuable tool that allows the farmer to be more in control of resources. Once you know how much money there is to spend and have clarified priorities, it is easier to say no to unexpected expenses as they pop up. I have also found that when spending temptations come along, having a budget helps to focus spending on items that assist farm productivity and avoid extraneous expenses.

When making spending decisions, I always ask whether the item is an investment in the future productivity of the farm, or just something that I want. Since farm funds are limited, needs come before wants. For example, I always wanted a decorative windmill for the farm, and was tempted to purchase one on several occasions. But, I knew that we needed to repair and expand our irrigation system, and that I had a very specific and limited amount of money to put into the farm each month. Having a budget helped me to prioritize fixing the irrigation, which would benefit the farm financially in the long-run, and delay buying the decorative windmill until necessary infrastructure was in place. I was able to put off my desire for a windmill joyfully, not begrudgingly, with the hope that the irrigation would help to increase farm yields and translate into more money in the budget down the road. By the way, I eventually did buy my windmill, and was able to enjoy it thoroughly, knowing that I had not ignored more necessary expenses in order to purchase it.

QR 9.4

KEEPING RECORDS: Food safety is a hot topic today. The demand for fresh, local food is increasing, but so is the demand for more and more government regulation. And, unfortunately, lawsuits are on the rise and anyone who runs an organization can be sued. Even if you are careful to ensure that your property is safe for visitors and your food isn't going to make anyone sick, you are still at risk. This should not scare you or discourage you from farming, as there are measures that you can take to protect yourself and your property.

The first line of defense is a safety plan coupled with a good record-keeping system. Good records not only help your farm to be more profitable, providing clues to where the farm is succeeding, and where its management can be improved; they also protect you legally.

At The Micro Farm Project, one of the items that we sell to the public is our farm eggs. Before we ever started selling them, we researched best production and handling practices and put a plan in place. Part of our research included taking a free Good Agricultural Practices (GAP) and Good Handling Practices (GHP) course offered by the United States Department of Agriculture. You can learn more about the program at ams.usda.gov/services/auditing/gap-ghp.

The course taught us what we needed to know in order to ensure that our farm products were healthy for consumption. I also consulted with a representative from Farm Bureau Financial Services, who gave me tips on how to make my property safe for visitors. Using this information, we came up with a safety plan for The Micro Farm Project.

But a plan, in itself, is not enough. In the event that someone would claim that one of our eggs made them ill, we need records to prove that we are following safe handling practices on a daily basis. This is where our egg production chart comes in handy. I created the chart on my computer, printed it and posted it near the coop with a pen attached so that it is very easy to fill out. Not only does it show what time and how many eggs we collect twice daily, but it also provides a record of when food, water and bedding are changed and how often we clean the coop and supplies. Filling the record out daily keeps us accountable and provides peace of mind that we are doing everything in our power to provide the freshest, healthiest eggs possible. And, in the event that someone would attempt to sue the farm, we have records to show that we were not negligent. Read more about our record-keeping system at the link in Figure QR 9.5.

INSURANCE: Good records go hand in glove with liability insurance, which is worth the cost for the peace of mind that proper coverage imparts. Even if good records are kept that show that the farm was responsible and not negligent, a lot of money can be spent to prove it. And even under the best of conditions, accidents do happen that can lead to a lawsuit. A good insurance policy will cover those legal expenses. Consult with your insurance agent concerning which coverages and how much coverage you will need. Some insurance companies will not insure a home if it is being used for agricultural purposes, so you may need to do some checking around. Additionally, if you partner with another organization for events, always ask for an *additionally insured endorsement* or rider that provides coverage for you on the partner's policy during the course of the event at no cost to you. Since every partnership and event are different, your insurance agent can advise you on how to structure insurance coverages with partner organizations to meet your particular needs.

QR 9.6

Discover more about planning a farm at the link in Figure QR 9.6.

SUSTAINABILITY STRATEGIES FOR LIVESTOCK AND POULTRY

GIVE EVERYONE A JOB

In previous chapters, we learned that livestock and poultry can perform numerous functions on the farm. Animals that are raised for meat and milk can also do many things to help out. Ducks can eliminate snails and slugs; goats can clear brush; chickens can turn compost and fertilize fallow gardens; fish can fertilize plants, and so forth. Pets can also perform useful functions, besides the natural enjoyment of having their presence on the farm. Dogs can discourage predators and provide companionship to livestock and cats can eliminate rodents. By employing all of the living elements of your farm, productivity can increase exponentially. Your farm will run more efficiently, with fewer financial and energy costs. Best of all, the animals will be happier and healthier, doing what they were created to do.

EXPLORE BUSINESS OPPORTUNITIES

Animals present many business opportunities, some of which are obvious. Eggs, milk and meat are probably the most common backyard products. Other opportunities are more obscure, but this means that they also have less competition. A creative farmer can find all sorts of ways to earn income. Many small income streams add up to a viable income and increase the stability of the farm.

At The Micro Farm Project, we marketed blown quail eggs and turkey feathers online to crafters. We sold and gave away bags of chicken compost. Chicks, point-of-lay hens, turkey poults and goat kids were sold to budding farmers. We traded bone broth, home-canned chicken, meats and homemade cheese for commodities produced by other farmers. It was a lot of fun to sell and trade products, and we gained a substantial amount of money and items that we needed for our farm.

I am often asked if we sheared our sheep to collect the wool. We did not shear nor sell wool, as our sheep were hair breeds that shed naturally in the spring. But farmers who raise wool animals can earn an income from the wool, especially if one learns to color it using natural dyes. I have come across farmers who produce garden meals, such as feather meal, bone meal and blood meal, which are made from waste products and are fantastic garden amendments. Crushed eggshells are also a possible commodity, sold for gardening and dietary purposes. There are so many opportunities for businesses, too many to mention here and more than I have fathomed. It can be extremely fun, rewarding and profitable to consider unusual business ideas, or ways to add value to the products that your farm produces.

Some business ideas are more profitable than others. And some farm activities are seasonal, providing occasional or periodic income. Having more than one income stream will improve the financial stability of the farm. And a substantial income can be produced by stacking small incomes from a variety of farm activities and products. Figure 9.3 shows tomato transplants sold by Farmyard in Phoenix, which earns a seasonal income and is one of many income streams for the farm.

9.3

BUSINESS STREAMS

DIVERSE PROFIT SOURCES IMPROVE FARM STABILITY

MULTIPLE INCOME STREAMS ADD UP TO A VIABLE LIVING AND FORM A RESILIENT FINANCIAL BASE

CO-OPS

Community Supported Agriculture often involves multiple small farms that band together to offer products to the community. But some products are more difficult to sell to the public, due to regulations by the Department of Agriculture and other government entities. For example, milk and meat sales are often highly regulated, making it nearly impossible for small farms to meet the requirements. One way around the regulations is to form a private co-op. For many years, a group of people in Arizona who wanted raw milk banded together to own a few cows. They shared the cost and the milk. The original owner of the cows provided the milking and distribution services. In return, she received funds from the members to support the care and feeding of the animals. Rules concerning cooperatives vary from state to state and county to county, so check with an agent from your local Department of Agriculture to learn the rules before starting a co-op.

CONCLUSION

Common sense dictates that a business whose expenses outweigh its returns will not be sustainable in the long run. But any business can improve its odds of success by reducing expenses and increasing its income. This is certainly true for backyard farms, whether they are commercial or private enterprises. By applying Permaculture principles and techniques, the farm can run more efficiently and with fewer expenses in terms of energy and finance. By creating multiple income streams, costs are offset and a reliable income can be produced. And for the hobby farmer, trading items can offset many of the expenses of farming, as well as household expenses. The goal is to receive a return that is greater than the effort and expense that farming requires. This will make farming worth the effort, much more satisfying, and ultimately more sustainable.

Sustainable farms contribute to the health and well-being of our properties and our families. But how can we extend this benefit outwards into our neighborhoods and towns? In the next chapter, we will explore the community benefits of urban farming, and how you can find your role in the local food movement.

FEATURED FARM
TERRA ROSA FERTILE FARMS

Like many Phoenix residents looking to escape the July heat, our family took a trip out of the valley in 2015 to northern Arizona. On the way, we made a small detour to Cornville to pay a visit to Susan Graves, one of my students in the Urban Farming program at Southwest Institute of Healing Arts (SWIHA). I was familiar with Susan's project, Terra Rosa Fertile Farms (TERFF), via pictures and design projects that she submitted during class, and was curious to see the project in person. Susan's design was of particular interest to me due to the complexity of building a farm on slope, and the serious thought that she was putting into arranging all the Permaculture pieces, from water harvesting, to soil health and biodiversity, and obtaining a yield. Although TERFF is not as urban as most of the farms that I am visiting for the City Farming project, as it is situated in a neighborhood of properties with acreage, Susan is applying strategies to make the most of her space that are applicable to smaller holdings and city lots, as well.

One of the striking features of the farm is the incline and the steep embankment that borders the property. Susan explains that she and her husband John originally viewed runoff water from the slope as problematic. "Our initial reaction was that when it rained, as quickly as we could, we would get all the water to leave the property because of flooding. Now we look at how we can actually slow the water down and use it before it leaves the property." Susan plans to use rain barrels and tanks to collect runoff from the roofs of the many structures on their land. She would also like to make better use of existing terraces to slow the flow of runoff water and to direct it, and she is figuring out how to implement a pond that is currently empty. One advantage of having such a steep grade is that water harvesting can be done passively, using gravity to direct water to containment areas and into the gardens.

The Grave's 9-acre farm was undeveloped when John purchased it twenty-two years ago. Susan explains how, "Several years ago, when I met my husband, we started maintaining the property, doing all the things that that you think are the right things to do until you learn that you're actually hurting your environment, such as killing native grasses and plants, as well as using poisonous chemicals."

"We started looking at a different approach on how we were treating the property; I wanted to be able to actually grow organically, embrace the native plants that were here and enjoy the natural water sources."

Susan and John started out simply wanting to grow food for themselves and for a few friends, and had no interest in growing commercially. "Raspberries are growing in the area near our neighbor's natural spring. In the garden a wide variety of herbs and vegetables are growing ... I have trellises near the house full of lavender, spearmint, and peppermint. This gives a great aroma near the house and is handy for salads and teas." Sue explains that, in addition to growing food, her initial motivation was to create an agriculturally and environmentally friendly environment by employing rainwater and graywater systems.

Recently, her vision has expanded.

"Currently, I'm taking a business class through Southwest Institute of Healing Arts, Urban Farming online program. It's called Seed to Sale. The idea that they give us is to go online and actually put together a business plan using the AgPlan.com website ... I ended up actually putting together a little business idea and even came up with a name for our farm and business; Terra Rosa Fertile Farms, in short TERFF, which actually fits in really well with some of the products that I want to be able to offer here. My tagline is "Promoting balance with Permaculture Principles, plant friendly products and services".

Sue is doing some interesting market research to find out what her customers want and need. "I'm advertising that 'We may have products and services that you are looking for. Please feel free to give us a call and if we don't have what you are looking for we may know someone who does.' When they see the advertisement and call, that helps me to learn what is marketable out there and what people are looking for. I get new ideas and bridge networking with other businesses and farmers."

Sue is really excited about bringing a particular product to market, and it's not anything that she grows on her farm! Instead, it's a resource that she uses on her farm, with great success, that she wants to share with others. "We've used it on our yard and in the garden. I have dense clay-based soil in a raspberry patch where I did not want to add anything too drastic to that soil. But I wanted to begin to amend the clay based soil's ability to provide what the raspberries needed. I started to think about organic soil amendments that we might be able to purchase. I used gypsum with a little improvement. Then a friend of ours, who has ownership in a humate mine, mentioned to me over a year ago the benefits of humate." Susan displays a handful of humate in Figure 9.4. Today, after amending the soil with the humate material, her raspberry patch is thriving.

"What I'm learning about humic acid is that it's a great neutralizer; that if you have clay soil that is difficult to work with, it not only will allow water to flow through it better, but it will also help to amend the soil and provides nutrients to the soil. On the other hand, if you have soil that was extremely sandy, it would also help to retain water and provide nutrients to the sandy soil."

Following the success of the raspberry patch, as well as improvements to her lawn and garden, the friend that originally recommended humate suggested that she might want to market the product in Arizona. "A year ago, we bought 2,500lb of humate, which comes in really big bags. It's a unique soil amendment, and this particular source of humate started eons ago as plant matter that decomposes and petrifies into humic acid. The humate mine takes the material from the ground and actually pulverizes it. It's not quite granular, but it's ground into small pieces that are workable so that you can add it to your soil." Susan discovered that there are no local sources for humate. She is excited to bring humate to market to provide a local source of the product and to save her customers the shipping fees that are associated with shipping small bags to Arizona from New Mexico. Susan also points out the great amounts of resources that she has available on her property, and has ideas to bring other soil amendments to market, as well.

When asked what she has learned that she would like to share with other farmers, Susan shows me the detailed plans and drawings that she has been preparing. "Before we consider more construction, we now have a different way that we're looking at the property and are doing some long-term planning. I think that's the biggest thing that I learned in the classes that I've been taking, all of that upfront planning that really needs to be considered before you just start digging in and wiping out natural areas. That's where we're at today, understanding that the key to having a great, productive and ecologically friendly farm is to do that plan upfront."

QR 9.7

Read the full interview with Susan and view pictures of her farm at the link in Figure QR 9.7.

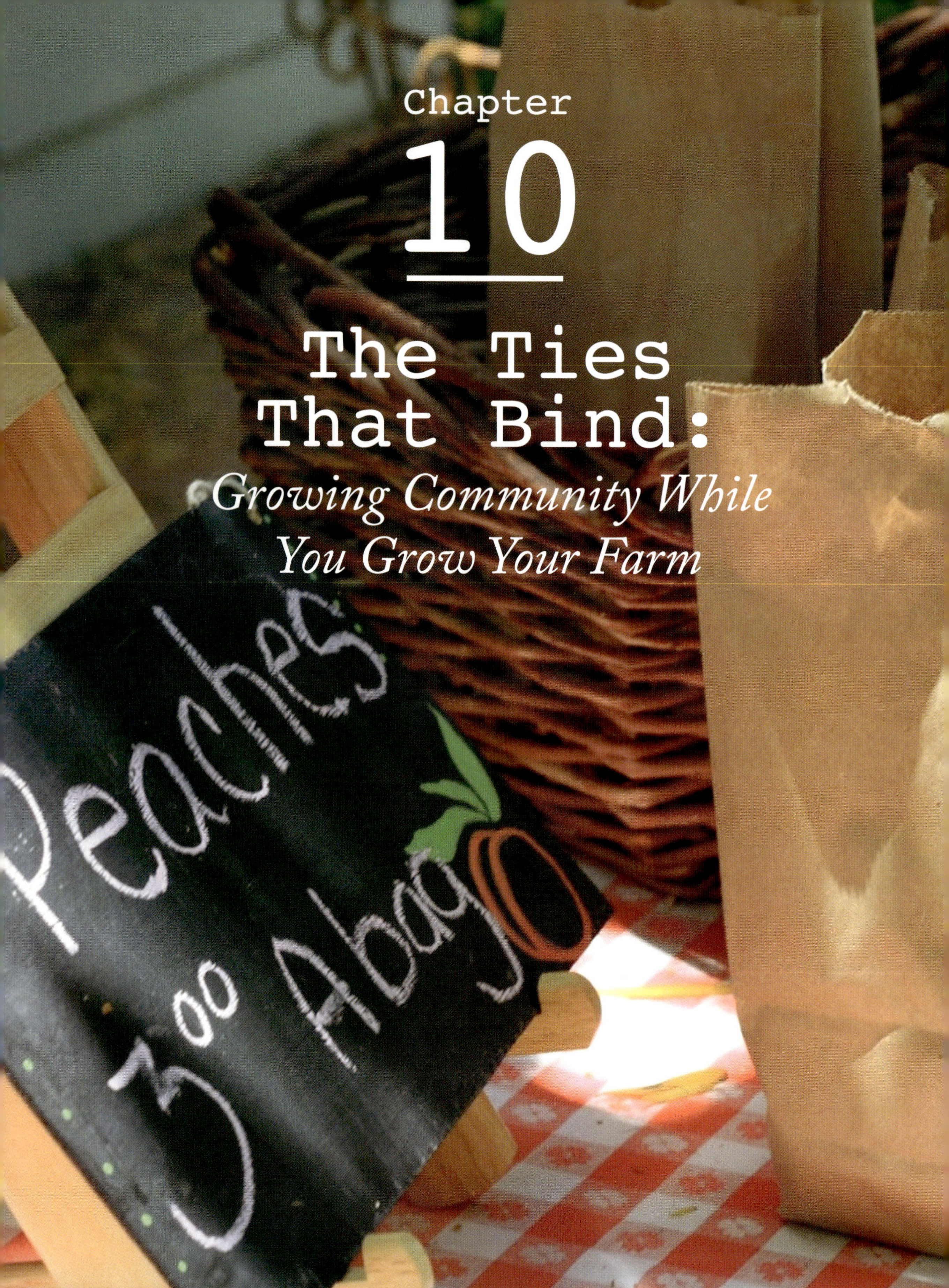

Chapter

10

The Ties That Bind:

Growing Community While You Grow Your Farm

In September 2016, we had to make a move. After living at our beloved farm for seventeen years, it became necessary to purchase a home that would accommodate our family plus my mother, who was recently widowed. I miss many things about our farm, from the pomegranate tree and the thriving gardens to the converted playhouse that served as our adorable chicken coop. But the thing that I miss most of all is the sense of community that we enjoyed with our neighbors and with the friends who joined us in our gardening adventures. The Micro Farm Project was perfectly located for building relationships, situated in a quiet cul-de-sac and surrounded by a group of friendly neighbors. And one of the best decisions that we ever made was to build garden beds in our front yard.

The front yard garden was at first intended only to expand my gardening space and provide a hands-on workshop for gardening students. My business partner, Greg Peterson, and I were teaching a course called *Growing Food: the Basics* at the time, and members of the class built and planted the beds as part of their training. Following the workshop, another group of students cultivated the beds with me over the course of several months to gain experience through an entire growing season. Eventually, this evolved into a garden club that met in our front yard on Monday evenings to grow food and enjoy each other's company. Figure 10.1 shows me with one of our outstanding garden club members.

10.1

COMMON GROUND

GARDENING BRINGS PEOPLE TOGETHER

PEOPLE FROM ALL WALKS OF LIFE WANT TO GROW FOOD AND EXPERIENCE COMMUNITY

Getting to know our long-term students and club members opened my eyes to the great need in our society for community gardens and garden clubs. People are hungry to get back in touch with nature and with the source of their food, and they are even hungrier for a sense of connectedness in an increasingly isolating culture. I also discovered, to my delight, that people of all walks of life can come together over growing food. In the garden, relationships thrive amongst liberals and conservatives, well-off and financially struggling, and all ethnicities and religions. Though we sometimes disagree about how to accomplish the goal, we all want better food, better communities and a better world for our families.

The urban farm has stretched me and my family in ways that we never could have imagined. When I first got interested in raising chickens, I received education and lots of help from a woman who had opposing religious views. We acquired goats and became good friends with a couple who held political views that were opposite to ours. And we visited many gardens and farms that were located in neighborhoods in which we otherwise could not have set foot. Quite auspiciously, I met my wonderful business partner, a man with whom I otherwise never would have crossed paths, at a tour of his urban farm. In every situation, we found common ground.

While we explored urban farming, we learned much about ourselves and other people. I don't know any other situation in which we could have become friends with some of the people whom we now hold dear. And what I appreciate the most is that, in a world full of increasing division, we are truly friends. As such, we are able to discuss all sorts of topics without the rancor that characterizes the current national conversation. I no longer hold my own opinions dear, and have learned that I can change my mind. I believe that this is a common experience in the gardening and urban farming world, as we all learn to live and to grow together.

As I become more involved in teaching gardening and farming classes, I also realize that people are concerned about a wide range of food issues. They want to support local and natural food production, as well as to produce some of their own food. They want to make a positive difference in the arenas of food and the environment. But they often do not know how to get involved. In this chapter, we will explore a few of the many ways to jump into the scene, to find your niche in the food revolution and make some friends along the way.

COMMUNITY GARDENS

Community gardens are popping up in neighborhoods all over the country. On a recent trip to New York City, I was surprised to find dozens of gardens tucked away in neighborhoods, as well as highly visible community plots, such as the Battery Urban Farm at Battery Park. Some of these farms are private, requiring membership, and are locked when members are not present. Some are open to the public, allowing the community to enjoy the garden whether or not they participate in cultivating it. A few have a mission, such as the Hell's Kitchen Farm, located on the Metro Baptist Church rooftop, that provides fresh produce to the Rauschenbush Metro Ministries Food Pantry. This diversity of garden types is not unique to Manhattan, but is becoming common in cities across the United States. To get involved, visit the garden locator on the American Community Gardening Association website at communitygarden.org/find-a-garden/. If you want to learn to grow food for your family but don't have room to cultivate vegetables on your own property, look for a garden that leases plots to members, like the one shown in Figure 10.2. If your goal is to learn to grow food while donating your time to help someone else, search for a garden with a mission. In either case, you will learn a lot from other gardeners and likely make some like-minded friends.

10.2

QR 10.1

SCHOOL GARDENS

School plots are becoming more prevalent as people's interest grows in teaching kids how food is produced, as well as giving them a hands-on experience in the garden. Whether or not you are a parent, school gardens are often looking for volunteers from the community. In order to become a volunteer, you may have to submit to a background check and fingerprinting, both of which help to ensure the safety of the children. But if you can jump through the hoops, gardening with children can be an extremely rewarding experience. Learn more about school gardens at the link in Figure QR 10.1.

4-H AND FUTURE FARMERS OF AMERICA (FFA)

4-H is a youth development organization overseen by the University Cooperative Extension system. It offers hands-on experiences for kids in a variety of programs, including agricultural projects. 4-H is always looking for mentors and volunteers. Find a program that interests you at 4-h.org. In exchange for volunteering your time, you can learn a lot about plants and animal husbandry alongside the children.

The National FFA Organization is similar to 4-H in its emphasis on youth development. It is, however, a private organization, not sponsored by the Cooperative Extension system. It provides agricultural classes, as well as hands-on agricultural experiences and entrepreneurship training. Many chapters are school sponsored. Adults are needed to help deliver programs to young people. Find a program at ffa.org.

SEED LIBRARIES AND COMMUNITY EXCHANGES

See section Seed Libraries and Exchanges in Chapter 7: Propagation

COMMUNITY EDUCATION

Many organizations offer free and low-cost gardening and urban farming education. In Phoenix, local horticultural nurseries advertise gardening classes. Public libraries and city programs offer gardening and irrigation information, and classes are generally posted on local Parks and Recreation websites. Local enthusiast groups, such as Trees Matter (formerly Valley Permaculture Alliance), hold regular informational talks and courses. Groups can be found online, particularly on social media and meet up sites. To find classes in your area, check your local library and community college events pages, city websites, Facebook, MeetUp.com, farmer's markets and nursery websites. Figure 10.3 shows a class that I am teaching at Uptown Grower's Market in Phoenix.

10.3

COMMUNITY EDUCATION

FREE OR LOW-COST CLASSES MAY BE OFFERED IN YOUR AREA

CHECK YOUR CITY'S PARKS & RECREATION WEBSITE, LIBRARIES, EXTENSION SERVICES AND COMMUNITY COLLEGES FOR LISTINGS

VOLUNTEERING AT A FARM

One of the best ways to learn, while also supporting local farmers, is to volunteer to help out. You may have to fill out an application and sign a waiver releasing the farm from claims should you be injured while on the premises. Many farms allow volunteers to sign up on their website. Some organizations require training before you are released to work. Donating your time to a farm can be a highly rewarding experience, offering education that is hard to come by any other way, as well as camaraderie and the satisfaction of supporting local food production. Our family spent time working at Lucky Nickel Ranch in Eloy, Arizona. Although it is a large farm, very different from our small city farm, we gleaned a tremendous amount of information from farmer Michael McKenzie. Learn more about volunteering at a farm at wikihow.com/Volunteer-on-an-Organic-Farm.

FOOD HUBS

Food Hubs are organizations that pool resources to help farmers distribute and market their products. Farmers markets and Community Supported Agriculture can be considered hubs, and these types of organizations often need volunteers. When you donate your time, you play an important role in supporting local food. In return, you will likely rub shoulders with farmers and food producers to whom you can turn should you have questions or need assistance in starting your own urban farm. Find food hubs by doing an internet search using the search terms *food hub* and the name of your city. You can also find food hubs at the National Good Food Network website, located at ngfn.org/resources/food-hubs/food-hubs#section-13.

PERMACULTURE GROUPS

Permaculture enthusiast groups are becoming more prevalent, existing in nearly every major city in the United States. These groups vary, from informal gatherings to structured organizations that promote permaculture education and perform projects. Getting involved with groups in your area will give you the opportunity to learn first-hand about permaculture principles and techniques from people who are practicing them. Having a group of knowledgeable people off of whom you can bounce ideas is invaluable to the process of designing and implementing your urban farm plan. You can find groups by performing an internet search. Groups can also be found on Facebook and on the MeetUp.com website at meetup.com/topics/permaculture/.

FOOD COALITIONS

Food coalitions are organizations that support local food systems. They often consist of think tanks, groups of people of diverse backgrounds who consider the issues and come up with ideas to improve the food system collaboratively in their area. Many also organize committees to carry out the ideas formulated by the group. The focus of a food coalition may include supporting farmers, improving access to fresh and local foods, and lobbying government officials on issues relevant to food production and distribution. Generally, expertise is not a prerequisite for joining a food coalition, just a willingness to support and improve your local food system. Most coalition websites can be found online and contain information about how you can get involved.

MASTER GARDENERS

I have referred to the Master Gardeners often throughout the pages of this book, and I would be remiss if I did not highlight them here. Master Gardener programs offer in-depth gardening information, including food production, that is specific to the region in which each individual class is held. If you desire to become a proficient gardener, the coursework and hands-on training are invaluable. Additionally, Master Gardener groups are service organizations that will allow you to give back to your community while making friends with other gardening enthusiasts. Figure 10.4 is of me hosting an educational table on a Master Gardener garden tour. Scholarships are available for some programs. Find a group near you on the American Horticultural Society website at ahs.org/gardening-resources/master-gardeners.

10.4

BEGINNING FARMER AND RANCHER PROGRAMS

If you are interested in running your farm as a business, the USDA provides funding to Cooperative Extensions and other organizations that provide education for new and prospective farmers and ranchers. Programs are designed to equip new farmers with agricultural skills and business tools that will increase the likelihood of profitability. Lewis and I took advantage of the Beginning Farmer program, as well as the Master Farmer course, offered through our local agricultural extension office. USDA

subsidies made these courses very affordable. We learned a great deal of useful information and developed relationships with several other farmers who live in our area. Use internet search terms *beginning farmers* and the name of your state to discover programs near you. You can also find Beginning Farmer training programs by contacting your local University Cooperative Extension or at BeginningFarmers.org/beginning-farmer-training-programs/. Lewis and I are in the back row of the photo in Figure 10.5, receiving our Master Farmer certificates.

10.5

Lewis & Kari graduate from a Beginning Farmer course funded by USDA

NEW FARMERS

HTTPS://NEWFARMERS.USDA.GOV/

FIND THE SUPPORT, EDUCATION, FUNDING & RESOURCES YOU NEED USING THE USDA NEW FARMERS DISCOVERY TOOL

COLLEGE LEVEL URBAN FARMING PROGRAMS

If you are interested in a degree or certificate in sustainable or urban agriculture, you are in luck. Colleges and universities have long offered agricultural degree programs focused on industrial agriculture. But more and more urban and sustainable farming programs are beginning to take shape. In Arizona, the Center for Urban Agriculture at Mesa Community College offers courses in aquaponics, greenhouse science, organic farming and meat processing. The center also offers free workshops and talks on urban agriculture to the public. Check with your local college or university to find a program near you.

THE DIRT

on farming

Academic Programs in Agriculture

CITYFARMINGBOOK.COM/DEGREE-PROGRAMS/

GARDEN CLUBS

One of the best experiences that I have had since becoming a gardener was joining the Washington Garden Club in Phoenix. I so appreciated the camaraderie and wisdom that I received from members who had been gardeners much longer than I. Meetings consisted of educational talks, tours, parties and service projects. Find a club near you on the National Garden Clubs website at gardenclub.org/clubs/join-a-club.aspx.

START SOMETHING NEW

My business partner, Greg, and I have created and hosted many gardening events, ranging from small classes to large summits. We are always willing to try something new. Our events have met with varying degrees of success, some having great attendance and others falling flat. In 2014, we put our heads together with Bill McDorman and Belle Starr of Rocky Mountain Seed Alliance. We were concerned about the lack of accessibility to locally adapted seed varieties, as well as the shortage of seed banks in Phoenix. In order to spur local gardening and seed-saving efforts, we created The Great American Seed Up, an event at which the public had access to diverse varieties of seeds at bulk prices, as well as seed saving and growing education. The event logo is shown in Figure 10.6. The Seed Up was a success and we have repeated it several

10.6

MAKE A DIFFERENCE

DON'T BE AFRAID TO START SOMETHING NEW!

OPPORTUNITIES IN URBAN AGRICULTURE ARE LIMITLESS AND THE NEED IS GREAT. IF YOU HAVE A GOOD IDEA, PURSUE IT

times. The event is fun and energetic, attracting new attendees every time it is hosted. We are now receiving requests to hold similar events in other cities and states across the country. You can learn more about the Seed Up at GreatAmericanSeedUp.org. And I encourage everyone to try something new. If you think you have a great idea, pursue it! Whether you reach five people or five hundred, you can make a positive difference in your community.

CONCLUSION

One of the benefits of urban farming that is often promoted is *self-sufficiency*. To some extent, I agree with this concept, if the focus is on being independent of the industrial food system. But no man is an island. Attempts to become completely self-sufficient are next to impossible in our society and, perhaps, misguided. The urban farming experience is so much richer when we are i*nterdependent*, sharing information and tools, trading labor and goods. Cooperation makes the urban farming experience easier; and camaraderie makes it fun! I encourage you to get involved in the local food community in whatever capacity interests you. Together, we can change the food system in our communities and make the world a better place.

FEATURED FARM

THE SOCIETY OF SAINT VINCENT DE PAUL PHOENIX URBAN FARM

The Society of Saint Vincent de Paul of Phoenix is an outstanding organization whose mission is to feed, clothe, house and heal our neighbors in need. The society serves 42,000 meals per day, which are supplemented with fresh vegetables from the Saint Vincent de Paul Urban Farm. I had the pleasure of interviewing Tony, one of the founders of the farm, who shared with me how the farm is feeding and healing people, and setting up mutually beneficial relationships among the farm, its clients and key players in the community.

I asked Tony about the purpose of the farm, and I received a surprising answer. "Healing starts with your fork, not at the doctor's or the hospital. Really, what you're putting in your body, it's going to be the most important part to living a healthy life in my opinion. And so we're really able to shift from being homeless and maybe getting something from a convenience store, whether it's a hot dog or some corn chips or whatever it might be, to a kale salad. You just are going to instantly feel better. That food is made to be eaten, not store it in a package."

"And as far as the healing goes, you also have people who come out here and volunteer. And I imagine that that's really healing the people, as well, to work in the garden. The garden is incredibly therapeutic." Tony told me that at one of their projects, they have discovered that 80% of the people who spent more than thirty days in the garden volunteering did not return to use the society's services. Instead, they found and were able to keep housing and a job. Tony says, "I think there's a lot of value to caring for something."

Saint Vincent de Paul also utilizes the farm space for special needs groups that come in from different middle schools and high schools, which have really responded positively to working in the garden. And the garden also has a regular volunteer flow of people who keep coming back to help out. "I mean, it's the middle of July in Phoenix and we have people that are spending their mornings out here with us for free," Tony laughs. "They're out here on their own and ... I think that we kind of have nature deficit disorder here in the

United States, especially," Tony explains that getting in touch with nature and cultivating the earth "bring a lot more happiness than texting or iPads". In Figure 10.7, Tony displays some of the bounty that the volunteers are growing.

MUTUALLY BENEFICIAL

TONY THINKS CREATIVELY TO CREATE WIN-WIN RELATIONSHIPS FOR THE FARM, SAINT VINCENT DE PAUL'S CLIENTS AND COMMUNITY PARTNERS

10.7

As a non-profit agency, the farm is dependent upon donations and volunteers to make it work. But Tony is thinking out of the box to set up win-win relationships with the community. As an example, he has plans to create a Food Truck club in which food truck entrepreneurs could grow vegetables for their businesses at the farm in exchange for financial donations and serving a certain number of free meals to the homeless each week. The trucks would benefit from the space to grow fresh produce and from Tony's expertise to help them out. Their customers would receive healthier and tastier food. The farm would be further supported by the donations. And the neighborhood would benefit from an increase in the number of healthy meals being served to those in need. Tony is thinking creatively about how he can keep the farm out of the red, with integrity, while keeping the focus on fulfilling the society's mission.

THE DIRT

on farming

QR 10.3

Saint Vincent de Paul Urban Farm Full Interview

CITYFARMINGBOOK.COM/SAINT-VINCENT-DE-PAUL/

Tony's advice to people who want to grow food or start something new in their community is to "just start. Just do it and look and research and learn. And I think the main thing, whether it's a small business or a garden or whatever you've got in your head, go and just do it. And you don't have to do it to the 11th degree. Start small and move from there."

Read the full interview with Tony at the link in Figure QR 10.3.

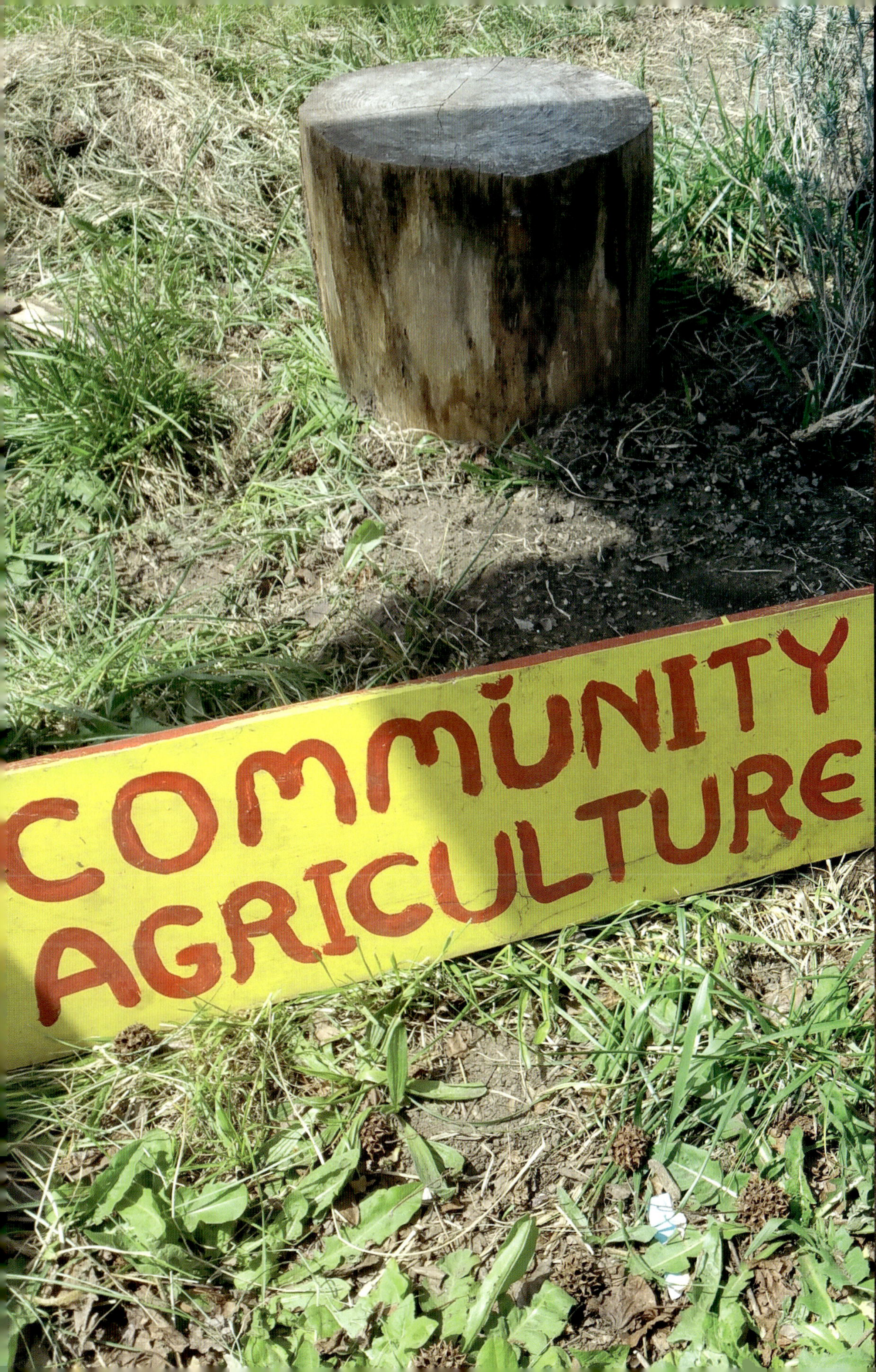
COMMUNITY
AGRICULTURE

Chapter

11

We End At The Beginning

If you are in the process of starting a farm, you are on the precipice of a great adventure. You are taking the first important steps toward fulfilling your dreams and creating the quality of life that you desire. Perhaps you long for a stronger sense of community, or you feel urgency to provide healthy food for your family and neighbors. Maybe you want to farm for the sheer pleasure of it, nurturing the land, growing vegetables and getting in touch with nature.

Whatever your goals, they may be thwarted if you get in over your head and lack the energy or manpower to carry them out. Exuberance at the outset may tempt you to want to do many things at once. Although it is a great idea to diversify your farm and bolster its profitability with multiple income streams, it is generally advisable to start with one or two elements and get really good at them before adding more to your operation. This will circumvent the potential for burn-out and will allow you time to build your community of workers and volunteers to assist you in accomplishing your goals.

Your urban farm design will be one-of-a-kind. Even when applying universal Permaculture design principles, every property or site presents its own unique character, challenges and needs. Using the guidelines in the *INVENTING YOUR FARM* sections of this book will help you to begin the process of assembling the framework of your farm piece-by-piece. Your plan will likely evolve over time, from the initial design through the implementation of the project. With a plan, more is accomplished than without one, even if the path changes along the way!

Here are some important points to consider in our final *INVENTING YOUR FARM* exercise:

DRAW IT OUT: The first *INVENTING YOUR FARM* section of Chapter 2 suggested taking the time to observe your property and draw an aerial view map of its features. This is the beginning of your base plan map. Use tracing paper to overlay objects onto your base plan, such as the future locations of gardens and livestock pens, potential water harvesting structures, a compost bin, or any other elements that you would like to include. Use a pencil so that you can erase and move things around as you think the plan through carefully. Don't worry if you have some uncertainty. Discover more about drawing a base map at Milkwood.net/2015/11/09/permaculture-design-process-2-making-a-base-map. My own design is shown in Figure 11.1.

11.1

PREPARE A BASE PLAN

USE GOOGLE EARTH TO PRINT A MAP OF YOUR PROPERTY

TRACING PAPER AND PENCILS CAN ASSIST IN THE DESIGN PROCESS. USE THEM TO OVERLAY AND ARRANGE FARM ELEMENTS ON THE MAP

CLARIFY YOUR MISSION: Grab your notebook and put the mission of your farm in writing. The mission is the reason for the farm's existence. Understanding and clarifying your own mission will help you to focus on what is really important to you and avoid distractions that will come your way. The mission and the design will be shaped by the personal goals you have for your property and it will reflect the lifestyle to which you aspire. With this in mind, ensure that all the members who will be sharing the space have input in the mission and design. By putting heads together, you will not only come up with a better plan, but you will also improve its chances for success. After all, the plan cannot succeed if all members are not motivated to invest their resources of time, finances, labor and mental energy into working the farm and fulfilling the mission for which it was started. Be realistic about your goals and aspirations, keeping these human resources in mind so that you don't exceed your labor capacity. The farm must be sustainable for the people who live and work there. If all goes well and more resources are added, you can expand your vision and the design.

RETHINK EXISTING STRUCTURES AND MATERIALS: What buildings, gardens, landforms and other objects are already available to you on the property? Can structures be utilized or repurposed to suit your needs? Are there any building materials available, or can you access them readily? As a Permaculture designer, the goal is to create a farm with easy flow that makes work simpler and more productive. If existing materials can be incorporated into the design without breaking the flow, it makes sense economically and environmentally to use them. If, however, they are obstructions that do not fit within the framework of your design or will require a lot

of upkeep, consider how to remove or repurpose them instead. For example, broken down fences that will need constant repair could be turned into garden trellises, shelves or decorative pathway markers. Replacing them with sturdy fences that will need less maintenance comes with an up-front cost, but the effort will save the farmer energy, worry and money in the long run.

SECURITY: Put secure fences and housing in place to protect animals and to keep their range to a minimum. This will help you to manage your livestock more easily. It will also give new plantings time to reach maturity and reduce potential damage by poultry and livestock. If predators are a potential issue, make certain that your fences are predator-proof. Urban farmers may not have trouble with dangerous predators, but strong fences will prevent your animals from getting loose and perhaps doing damage in your neighborhood.

Fences and sheds also protect your valuables so that you will not spend time, money and energy replacing or repairing items that have been stolen, misplaced or vandalized. While most people will respect and appreciate your farm, removing the temptation from those who would do damage increases security and maintains peace in your operation. It is important for farms to be inviting to the public, but it is beneficial to have secure areas that only you and your workers are able to access.

If you can find used or recycled fences, make sure they are sound and that the hardware is working correctly. Gates and latches that don't work properly rob the farmer of time and energy, and make farming more difficult. On the other hand, properly functioning equipment makes work easier and increases farm productivity and flow.

KNOW THE RULES: In many ways, urban farms break the unwritten rules of our current food system. But, in order to secure your property and avoid an unwelcome visit by the local authorities, it is important to develop a design that does not blatantly violate planning statutes and guidelines. Make a call or a personal visit to your local government authority and a talk to a town planner concerning your design goals. They are usually pleasant advisors whose job it is to help you to stay within the bounds of the legal requirements. It is a good idea to maintain good working relationships with them.

Properties are subject to zoning guidelines and restrictions that govern how you can use and arrange them. For example, there are rules governing how close permanent structures can be placed near the property line to avoid encroaching on easements.

Certain types and heights of buildings may be restricted. Water harvesting may be regulated. In some municipalities, food gardens cannot be grown in visible areas of the property and certain types of plants may not be allowed at all. Some cities welcome chickens and livestock; others do not. Check the zoning and associated restrictions that apply to your property early in the process.

Sometimes local laws might seem archaic and counter-productive in a society where ideas about food production are evolving. As an urban farmer, you will have a platform on which to initiate change. While following the rules to the best of your ability as they are currently, you can simultaneously actively educate your community and lobby local lawmakers for adjustments in favor of urban farming, water harvesting, environmental concerns and other issues that affect your operation and your community.

TAKE INVENTORY OF RESOURCES: You can minimize costs and increase the sustainability of your plan by assessing and utilizing the resources that you already have available to you. Everything on your site is a potential resource, from rocks and weeds to wildlife and landforms. Here are some of the potential resources to consider:

- Finances: Start-up ventures carry risk, so it is wise to start within one's financial means. In my opinion, it is advisable to avoid borrowing and the resulting interest charges, which can add stress to any project from both a budgetary and a mental/emotional standpoint. The availability of cash to fund material purchases and professional services will greatly influence how and when various aspects of your plan are implemented. I have discovered that there is often a creative, thrifty way around expensive farm solutions. Thoughtfully considering lower cost options and solutions will help you to establish your design more quickly and economically.
- Labor: How much strength and physical ability do you have? And how energetic are your team members and volunteer workers? Though the ultimate goal is to create a low-maintenance Permaculture system, initial establishment of the system may require a considerable amount of work. Hiring outside labor adds to start-up costs. Plan your design to stay within the available human and financial resources.
- Skills: How much of the skilled labor will you be able to accomplish yourself or within your social and volunteer networks? Will it be possible to barter goods or your own talents for skilled labor? Or will you need to hire and pay for the expertise needed to accomplish your plan? Often, low-cost, low-tech

solutions are available. The internet is replete with tutorials for systems that anyone can implement and maintain themselves. But high-tech solutions are sometimes appropriate, as in the case of solar energy. Perhaps someone in your network can help you with know-how that you do not personally possess. If not, ask for recommendations and hire a trusted professional. It may cost you more up-front to hire a qualified professional, but the benefits of having the job done correctly and efficiently often outweigh the initial cost.

Additionally, find out what skills exist in your farm team and social network. Talk to workers, friends and neighbors, making notes of who is willing to help out and what skills they possess so that you can call upon them as needed. Be mindful of their needs and desires, too, so that you can reciprocate, providing benefits in return to those who offer you their help. Mutually beneficial relationships build community, and some people will be interested in helping you out simply for the enjoyment of spending time on the farm and learning about what you are doing. Others may appreciate the offer of fresh produce or grateful acknowledgement of their part in helping to build your operation. Your generosity will invite the generosity of others.

- Treasure hunting: Even if you are not yet ready to implement your design, you can begin to collect helpful resources. Make a list of items that you anticipate needing, such as plant pots and tools, building materials, shelves and so forth. If you have a storage area in which to keep a stockpile, look out for free and inexpensive resources at recycling centers, dumps, thrift stores, bulk trash pickup days, classifieds advertisements, public notice boards and amongst your social network. If you do not have a place to physically store such items, keep a record of places that sell or give them away so that you can check back later when you need them.

WATER: Establish a connection to water, whether it is from a municipal water source, a natural source or a well. As Greg Peterson of UrbanFarm.org recommends, *plant the water before you plant the garden*. Make plans to establish water self-reliance (as much as possible,) which is foundational to self-sufficiency and productivity. Earthworks should be accomplished early on if you are planning to do contouring or dig formations to slow down and collect rainwater. Additionally, you may consider setting up tanks to collect rainwater and establishing graywater systems. Look into local water conservation societies or Permaculture groups that may be able to assist you with low-cost solutions and implement them as your budget and cash flow permit. Figure 11.2 shows the drip irrigation system installed at The Simple Farm in Phoenix.

11.2

ESTABLISH WATER CONNECTIONS

PLANT THE WATER BEFORE YOU PLANT THE GARDEN

IF SUPPLEMENTAL IRRIGATION WILL BE NECESSARY, INSTALL THE SYSTEM FIRST, PRIOR TO PLANTING

START YOUR NURSERY: Biodiversity is critical to the design of a good Permaculture system because any system's productivity is highly dependent upon the number of living elements it contains and the productive connections between them. As you begin to carry out your plan, start early on to propagate plant species that are well suited to your site. To do this efficiently and inexpensively, set up a plant nursery using the propagation techniques that we learned in Chapter 7. Start simple; growing just one or two of your own varieties will give you practice and contribute to the sustainability of your farm.

SOIL BUILDING: Hand-in-hand with establishing plant propagation systems, improving your soil should be a priority. Plantings cannot be successfully accomplished until erosion is controlled and the soil is prepared to meet the needs of the crops. Methods may include growing cover crops, establishing windbreaks, adding organic matter, introducing nitrogen fixing plants, and aerating to correct soil compaction. See Chapter 6 for specifics.

START AT HOME BASE: The highest priority when starting a sustainable farm is to develop the nucleus of your property – your home or base of operation, as well as the area immediately around it. This area is where you will locate elements that you want or need to visit often, such as a kitchen vegetable garden, a chicken coop, living and office spaces. It may also include spaces for dining, recreating and aesthetically pleasing areas in which to rest. Additionally, you may want to plant carefully chosen orchard trees that provide your favorite fruits or nuts and provide shade to your living area in the summer.

If your property is small, the entire area may be considered your home base. But if you have a larger property, establish your homestead or base of operation completely first, then spread out from there. Though it may be tempting to make improvements over the entire property simultaneously, focusing on your home base will conserve your energies and establish infrastructure to support the management of the rest of your property. It will also give you a place to rest and recharge from your efforts.

EXPAND MINDFULLY: Once the home base is established, you can begin to develop the outer zones in stages. You may want to start with a few plantings of reliable varieties for sale, increasing the size and scope of your commercial beds as you gain farming experience. Develop a steady cash flow by staggering your plantings and growing varieties that mature at different times. If you are planning to bring livestock on to the farm, do so slowly, establishing efficient systems and strong connections amongst the elements in order to make the most of their contributions to the farm. Consider *value added* products, such as baked goods, jams and jellies, cheese, candies and other products to sell during seasons when produce is not being harvested. Check your local regulations concerning the sale of homemade products and potential outlets. Some things that you do will surely succeed and others may fail, but slow and mindful expansion will help you to avoid mistakes and build your farm on a solid foundation.

TAKE IT STEP-BY-STEP: Break down the implementation schedule of your design into stages consisting of small, simple steps. Write them out on paper and give each step a target date. Move through the steps one at a time, thoroughly completing and implementing one step before moving on to the next stage.

Start by putting basic infrastructure in place on your property. Establish access points, such as gates, paths and drives. Build shelter for yourself, if it is not already in place.

Barns, workshops, sheds and patios ought to be ready prior bringing animals, equipment and other valuables onto the property.

Sustainable farms do not happen instantaneously. However, with patience and strategic implementation of carefully planned design, your urban farm will develop increasing food security and energy self-sufficiency, as well as reliable yields of farm products with minimal inputs. The farm can become an asset and connection point for the community, as well as providing a nurturing lifestyle for the farmer. And it can yield a satisfying livelihood, providing delicious and nutritious products and useful, marketable services to the community.

CONCLUSION

Throughout the pages of this book, we have discussed Permaculture systems extensively. In contrast to industrial agriculture, your urban farm can go beyond sustainability, actually improving the environment and increasing its productivity and biodiversity. It is very exciting to consider that a simple urban farm can contribute to the renewal and restoration of the metro environment, providing fresh, local produce in areas that would otherwise rely on transported goods. This not only reduces dependence on oil associated with food transport, but also increases food security in the cities. Commercial urban farms tend to be market driven, meeting the needs and wants of the neighborhood and providing gainful employment to residents. City farms also serve to beautify marginal land and help to mitigate the harmful effects of pollution.

Additionally, urban farms are innovative, growing food in unique ways that make creative use of local resources. By applying the Permaculture Principles and sharing with your neighbors, your farm will contribute to human and ecological health and well-being. Perhaps most exciting of all, your farm can become a showcase of regenerative systems that will inspire others to go beyond sustainability within their own spheres of influence. In this way, the message of Permaculture will spread and more regenerative systems will be implemented. As this occurs, farm by farm and within homes and workplaces, our cities can be transformed into healthier, more inviting spaces where people are able to thrive for generations. In a very real sense, urban farms are saving the world, one city lot at a time.

FEATURED FARM
STONE HOE GARDENS

The very first farm that I visited for the City Farms project was Stone Hoe Gardens, a 2-acre, irrigated property in south Phoenix. Farmers Bruce and Tina Leadbetter have been friends of ours for a number of years, and we have joined forces on more than one occasion to accomplish farming tasks. Stone Hoe Gardens is a lively farm, with extensive vegetable gardens, fruit trees, Nubian goats, chickens and a handful of lucky dogs who roam the property freely.

I was excited to visit their operation and practice my interview skills on them, and we had a blast! I had a pretty good idea going into the visit about what they had going on at their farm, but I learned some new things, too!

The first question that I asked was about the name Stone Hoe Gardens. I was curious to know the history behind the moniker. Tina told me the story, "We have this wonderful property in the Phoenix area that is just south of the Salt River. This area has been inhabited by native peoples for a very, very, very long time. When the current neighborhood was developed back in the 1950s or '60s, they plowed the road in front of our farm and turned up thousands of stone hoes left behind by a Native American village that was in this area. And so it seems that there were farms here for hundreds of years before we came. We named the farm after this discovery and all of the stone hoes that were found in the road."

When asked how the farm got started, they tell me that Bruce grew up having farm fresh eggs. When they purchased the property, Bruce suggested getting chickens and Tina was game! As they related the story, Tina laughed at her initial reaction to the eggs – "You should've seen the suspicion with which I approached our new hens' eggs the very first time. I was almost afraid of them! I certainly don't know why, looking back now. I think it's laughable because I refer to eggs you can buy in the store as fake eggs because they don't look anything like our farm eggs."

Top Tips from Stone Hoe Gardens

- Shade: In our hot, low-desert climate, planting vegetables on the east side of trees so that they get morning sun and afternoon shade is helpful to keep gardens from burning up in the summertime heat. In areas that do not have protection to the west, shade cloth can be helpful. Shade cloth comes in different levels of UV protection, much like sunscreen. Be careful not to shade out production, however. Some plants (such as tender greens) may like 90% shade protection; others (such as tomatoes and peppers) will benefit more from 30% shade.
- Water: Although the Leadbetters have access to flood irrigation, they use it primarily for trees and grassy areas because it is only available twice per month in the summer. Veggie gardens need to be watered more often than that. To conserve water, they use drip irrigation for precise watering several times per week. When they researched drip irrigation systems, they discovered that an irrigation system called drip tape claims to be 90–95% efficient (in contrast to flood irrigation, which is only 70% efficient.) Drip tape is easy to install, but it is susceptible to rabbits and other wildlife that will chew through it to access the water. Tina recommends burying the drip tape underneath the soil to hide it from animals, and also to extend the life of the system by protecting it from harsh sunlight.
- Volunteers: The Leadbetters recount how they had not anticipated the amount of work required to run a farm. To reduce the load, they turned to volunteers. Tina says it's a win-win situation, "It's a really great trade-off to mitigate the amount of work that goes into a farm. And it's an opportunity for our volunteers to be able to see what works and what doesn't work, and how we try things. They also get a little bit of personal experience if it's something they plan to do in the future. Some of our volunteers are being very smart about learning this lifestyle before they try to set up their own system. Others volunteer because they can't have goats at home, so they have goats here. Or they can't grow a vegetable garden because they live in apartment, so they get that experience here. It's up to the volunteer. These arrangements have really worked out very well for both parties." Read the full interview to find out how they find their volunteers.
- Use Free Resources: Urban farms have many expenses. In order to reduce the financial burden and be environmentally friendly, Tina and Bruce

use the resources that they already have on the farm and look for free sources of off-farm materials. As an example, following election cycles, they collect abandoned political signs. These signs are made of durable materials, but are easy to cut and shape. The Leadbetters like the idea of turning a neighborhood eyesore into a resource for their farm, and they use the signs to provide shade protection to their livestock, as shown in Figure 11.3.

USE FREE RESOURCES

ABANDONED SIGNS MAKE A FENCE AT STONE HOE GARDENS

11.3

- Cooperate: Partnering with other farmers is one way that Bruce and Tina are able to expand their operation, gain knowledge, and share the workload. We have personally worked with them on poultry processing days to make a difficult job easier and more fun! They are cooperating with other farmers, as well. Tina is keen to have bees on her property, but is hesitant to jump in with both feet without learning about them first. Because of the size of her property, she will be providing space for a local apiary to place some of its hives. In return, she will learn hands-on how to care for them, right in her own backyard!

- Have fun! Bruce recounts how they started out selling vegetables to restaurants, which was great! However, trying to fill orders soon became burdensome and they knew that they could not sustain that sales model if it was not enjoyable to them. They modified their system, and started emailing their customer list with the things that were available on the farm each week. According to Bruce, "We decided to just pull back and have fun with it. Whatever we were growing at the time, we posted in our newsletter. And the customers started showing up. You know, I think the lack of that negative energy made the difference." They found that eggs and milk would sell quickly, and Bruce was able to apply some creativity in making cheese, jams and other specialty items.

Stone Hoe Gardens has had its share of successes, as well as things that didn't turn out as well as they had hoped. The Leadbetters consider everything to be an educational experience, and they are always learning, growing and adjusting their farm to fit their personalities, dreams and goals. Tina tells me with a twinkle in her eye, "We are not certain about how it's all going to work out, but I think that the one thing that we've settled on for sure is that we love knowing where our food comes from. And we love helping other people discover it, too!"

See pictures of Stone Hoe Garden and read the full interview with Bruce and Tina at the link in Figure QR 11.1.

Afterword

As the writing of this book draws to a close, I pause to reflect on the wild ride that our family and our farm has taken over the last two years. When 5M Publishing contacted me early in 2015 about the possibility of creating a book such as this, The Micro Farm Project was in its heyday, and I was elated to share it with the world. The gardens were lush, and people flocked to tours for the chance to see dairy goats, sheep, turkeys, quail, chickens and rabbits on a city lot. We took the opportunity of an audience to share what we had learned and how we were creating a more sustainable city farm by composting, using greywater, and applying the principles of Permaculture to the operation.

To us, urban farming had become part of our normal, everyday life. But to many of our neighbors, it was a novelty. My earnest desire and mission was (and still is) to make metro farms an ordinary, accepted and celebrated aspect of city life with large numbers participating. I was doing my part to reach out to potential farmers by teaching, hosting events, writing college curricula, opening the farm for tours and blogging at TheMicroFarmProject.com. Then, 5M Publishing offered the opportunity to write a book that would reach a new and wider audience, and I was all in!

Once the outline was completed and the publishing contract was in place, I began with gusto to gather research, visit farms and prepare the City Farming website. Although I was thoroughly enjoying the work, it was overwhelming at times to keep up with the book project, as well as my teaching and event schedule and staying on top of the daily tasks of farming. Fortunately, I had a lot of cheerleaders around me who offered encouragement and assistance. Lewis helped tremendously with farm tasks and traveled with me to distant farms. My girls stepped up and helped out whenever they could, and Greg took up slack with our business, GrowPHX. And, although he was ill and could not help out, my father's obvious pride and enthusiasm for the project fueled a fire under me to complete it with excellence.

In April 2016, our world changed. My father passed away, and we invited my mother to live with us. Unfortunately, we quickly discovered that the cost to remodel The Micro Farm Project was more than our budget would allow, and that moving to a home that would accommodate all of us would be more economical. Following an exhaustive (and exhausting!) search, the only home that we found that would accommodate all of us was in a neighborhood that would not welcome the farm in its current configuration. So, we set about finding homes for the animals and farmers who could use the enclosures, tools and other supplies that we would no longer need.

While we were very happy to add my mom to our household, it was an extremely difficult process, both physically and emotionally, to dispose of our farm. I miss the gardens, people and animals tremendously, especially the garden club members, Calamity Jane, the doe who was born on our property, and our friendly ram, Oliver. But, I take comfort in the idea that we helped out a lot of our farm friends who gratefully carried away valuable pieces of our life to put to use on their own special farms.

Life has a way of throwing curves at us. The best we can do is to roll with the changes and look for the opportunities that lie within the challenges. It would be easy for me to focus on what I *can't* have at my new property – no goats, no sheep, no large tours, and, possibly, no chickens. I am choosing instead to look at what I *do* have – gutters for water harvesting, a large patio for container gardens, a mature grapefruit tree, and a north-south exposure.

Most importantly of all, our family has the opportunity to design a new kind of farm, equipped with the knowledge that we have gained both from our experience and from information that I have gathered for this book. It will be more challenging to produce food on the new property, for certain. Lewis and I will have to try some new ways of doing things, perhaps delving into aquaponics and vertical gardening. I believe that we are up to the task, and look forward to the unfolding of our little farm with anticipation. But, most of all, we face the future with great expectation as the city farming movement grows, enriching our individual lives and our communities like no other pursuit.

References

Favell, D. J. (1998) A Comparison of the Vitamin C Content of Fresh and Frozen Vegetables. *Food Chemistry* 62.1, 59–64.

Shaw, Arch Wilkinson (1927). *The Magazine of Business* 52, 182.

The Encyclopedia of Earth (2014) Ecosystems. Available at: http://editors.eol.org/eoearth/wiki/Ecosystems. (accessed November 2016)

How Stuff Works, a Division of InfoSpace Holdings LLC (2008) How Low-VOC Paint Works. Available at: http://home.howstuffworks.com/home-improvement/construction/materials/low-voc-paint.htm. (accessed October 2016)

Global Issues (2014) Why Is Biodiversity Important? Who Cares? Available at www.globalissues.org/article/170/why-is-biodiversity-important-who-cares#WhyisBiodiversityImportant. (accessed November 2016)

Index

C

S

W

Y

Z